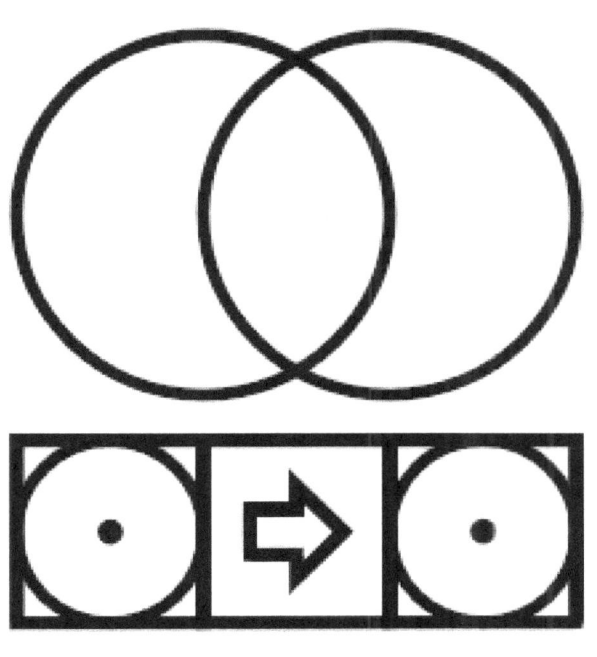

GEOMETRON

by Trash Robot

The Book of Geometron

Trash Robot

May 28, 2021

Contents

Contents	iii
List of Figures	v
1 Civilizations	1
2 Organic Media	16

3	Street Network	38
4	Servers	60
5	Scrolls	68
6	Feeds	86
7	Maps	97
8	Symbols	110
9	2d Web Graphics	134
10	Shapes and Fonts	163
11	Action Geometry	184
12	Printers	210
13	Geometron in 3d and Beyond	224
14	Magic	241

List of Figures   v

15 Full Stack Geometron   255

List of Figures

8.1	vesicapiscis	116
8.2	pentagram	118
8.3	hexagram	119
8.4	squareroottwo	120
8.5	scales	121
8.6	cursormovements	122
8.7	colors	123
8.8	uparrowspelling	124
8.9	cubes	126
9.1	keyboard	136
9.2	cursoredit	138
9.3	move	139
9.4	angles	140
9.5	scaleactions	141
9.6	basicdraw	142
9.7	colors	143
9.8	pathactions	143

9.9	bezieractions	144
9.10	bezierbracket	145
9.11	panzoom	146
9.12	flagactions	146
9.13	vesicapiscis	147
9.14	cursorscale	148
9.15	cursorangle	149
9.16	cursorsquare	150
9.17	cursorroot2	151
9.18	cursorgolden	152
9.19	cursorroot3	153
9.20	cursor3	154
9.21	cursor5	155
9.22	styleeditor	156
10.1	RLC	169
10.2	inductorloop	170
10.3	inductor	171
10.4	RCline	172
10.5	treeoflife	173
10.6	treeoflifespelling	174
10.7	estrogendiagram	175
10.8	estrogenspelling	176
10.9	hebrew	177
10.10	katakana	178
11.1	shapeset	192
11.2	rulers	193

List of Figures vii

11.3 protractor 194
11.4 penrose 195
11.5 stencil 196
11.6 pilogo 197
11.7 skeletrontripod 198
11.8 skeletron2 199
11.9 shook 200
11.10 tapesnake 201
11.11 artboxnet 202
11.12 artbox 203
11.13 artboxtop 204
11.14 pyramidnet 205
11.15 pyramid 206
11.16 bags 207
11.17 outfit 208
11.18 outfit 209

12.1 basicmovements 211
12.2 actions05xx 213
12.3 printerphoto 215
12.4 printerblockdiagram 216
12.5 trace 218
12.6 iconfeed 219
12.7 buildingwallrobot 220
12.8 eblblockdiagram 221
12.9 pendant 223

13.1 actions3d 226

13.2	shapebuilding1	227
13.3	shapebuilding2	228
13.4	whimsycastleturret	232
13.5	whimsycastlespelling	233
13.6	whimsycastle	234
13.7	robot0500	237
13.8	trashrobot3d	238
13.9	angles	240
14.1	pisigil	248
14.2	actiongeometrysigil	249
14.3	alchemy	251
14.4	sigilblank	252
14.5	setblank	253
14.6	setblank2	254

Chapter 1

Civilizations

Dig it up, set it on fire, and bury it. Our civilization is an ever-accelerating destructive flow of material from mine to landfill. This work is based on the idea that we can do better. Much better! We have dug such a vast array of useful minerals up in the last few hundred years and carried out such fantastic transformations on them into useful materials that we now have the opportunity to build a whole new civilization with a completely different structure than the one we presently inhabit.

Our current industrial system is a tripod resting on these three legs: money, mining and property. These ideas make up a philosophical framework for understanding and building our world which is failing at an ever-accelerating rate. In order to build a new civilization, we must study in detail the structure of the existing one.

Money as an idea is so integrated into our world view that it is very hard to even see what it is. At its philosophical core, the idea of money is that there is a property we call "value" which can be denoted by a number. This is considered so obviously true by defenders of the existing economic system that when challenged, they simply re-state repeatedly the basic ideas immediately assuming that any system not based on money is just the same set of ideas re-arranged. That is, people will argue that doing away with money will simply replace selling a things with a certain "value" denote by prices with trading things to which such a value can also be assigned, but just in a much more inconvenient way(barter). Similarly, the defenders of the idea of money will argue that "storing value" is an important task, again assuming that some kind of thing called "value" can be measured with numbers and that doing this all-important value storing will simply have to be done with vaults of metal if we dispense with money.

But this monetary way of thinking does not take into account the possibility that the value of a thing might multiply by replication or that value can be created from nothing. As long as everyone in society is exchanging value along this stream from landfill to mine, we can use the numbers we call "money" to roughly create an equivalent to the main type of value, which is physical material. We can measure how much gold or cobalt or salt we have, and money lets us transform value from one of these physical things to another, easily trading lithium for silver or oil for aluminum.

Furthermore, of course, there is labor. In the labor theory of value, again, money is used to assign some fixed value to a certain amount of "work" people do to produce a thing. One can superficially dispense with money but as long as we accept that value comes from how many hours an actual worker one does some type of task, we are just shuffling the details around but preserving the structure. But what of automation? And what of other forms of even more drastic increases in efficiency which are possible with information technology? Again, once we allow for full automation and many orders of magnitude increases in efficiency, we find that there is no way to create a value system using numbers which actually describes reality as we experience it in an information economy. Numerous band aids have been proposed, but if we project forward to more and more automation and further increases in efficiency we see that we have to re-evaluate the whole labor theory of "value" along with every other theory of value we hold as a truth today. The problem, again, is that using numbers to denote value simply will never be compatible with the new civilization we must now build in order to survive.

What happens when we create value from something we have an effectively infinite amount of but which we only have a finite need for? What if, for instance, I live near a landfill with a vast store of plastic and electronic trash, and I want to build a small factory which converts plastic trash into useful furniture for direct use near the material I have. Suppose the electronics trash has all the material I need to build this fully automated factory, and that the material needed to create so much great furniture that no one near me ever has to buy furniture again, that they can get custom high quality furniture which can be repaired indefinitely without even tapping too deeply into the vast store of plastic trash available directly in our community. This act of creation brings a thing of great value into being from nothing but information. In this situation, money fails. As long as we rely on money to denote and store value, everyone creating value from nothing has to either get someone in the money-creating business to create currency specifically for them or to do that themselves as they create.

But the more creative our new industrial processes are, the more catastrophically the money system fails. If I share the information on how to build the thing with the rest of the world, in principle the equivalent of trillions of dollars of value can be created from nothing.

This failure is not hypothetical. Creating value from nothing which can self-replicate freely is precisely what software does. Software produces things of great value with no material input needed at all, and is able to almost instantly replicate to the whole of humanity. If someone can create almost infinite value with no input of labor, energy or materials(after the initial creation process), what does that imply for the rest of the people? We are seeing this every day. Every city in the world right now is witnessing a violent takeover by the replicators. Those who can replicate their products infinitely for free include the media industry, marketing, software, finance, and all the numerous information based businesses which make up those business ecosystems. We are finding that at an ever-accelerating rate all the wealth in our society is being transferred from people who produce things that do not replicate, like physical goods or physical labor to the replicators.

The purpose of the work here is to show that there is another way. We can create an information based economy made up of self-replicating information which reproduces things of value which cannot be added up using numbers. This is not some crude replacement for money using barter, but rather a whole new approach to everything: we will have to build a new way of thinking about machines, society, mathematics, and philosophy. Just as the existing system is based at its deepest level on numbers, we propose basing our entire civilization on geometry. Geometron is a universal geometric language which we can use to express the self-replicating information we will need to build this new civilization.

The second leg of our tripod of civilization is property. As with money, the idea of property is very difficult to examine as it is so deeply woven into the fabric of everything we do. The world view we are taught in school and at home is that just as everything has a price that everything is owned by someone. In some cases that ownership might be the state or even some type of "common property" but all things are in some way property. In the dominant ideology of our time, air and water are property, the land is property, the genes in our DNA patented by the drug companies are property, and even these words I am typing are property.

Again, as with money, we have to examine how this idea is going to fail more and more catastrophically as we evolve into a civilization based on the replication of technology from trash rather than consumption of mined materials using labor. In many ways the purpose of property is to inhibit replication. In the case of intellectual property this is in fact its *only* purpose. But even for physical property like land, the whole idea is that if I "own" land the real purpose of that ownership is to make sure someone else does not own it.

The idea of property makes sense when we are all competing for resources we have to dig up out of the ground. If I dig up 1 ton of silver it means you can't and vice versa. We are all in competition to be the owner of that silver. If you are 1000 pounds of silver richer, I'm 1000 pounds of silver poorer. But we have to see that with an informational economy based on trash as the main input this is no longer the case.

If I consume a ton of trash from a landfill into useful things, we have to recognize that that ton of trash has a negative value, which changes completely what it means to use it to make things. If I consume 1000 pounds of oil to make a plastic part, that oil has a cost we now measure in money I have to spend to buy the oil. But if I consume that from a landfill, the cost is negative! Regardless of what I make from the plastic, simply having it be something useful at all is of value. In our existing consumer economy we all actually have to spend money to dispose of waste. So an economy built entirely on waste streams breaks the whole idea of property up. If I have a pile of trash on my land and you take it away that has value to me and also to you. We are no longer in competition in this relationship–the more trash you take the more benefit I get but also the more you get.

We must also recognize how an economy based on the replication of technology from trash using free information technology changes the incentives in regards to intellectual property. If I create a new technology now and release it into the world, the only way for me to make money on it is to retain some control, which is now expressed by means of the property system. But if I create a new technology from trash which does something useful, and it's intended to only be used locally, the value proposition changes. I use local materials to make a thing and directly benefit from it. When I share it with you, you also do that, but if you then improve upon it and release it back into the network, I can immediately benefit from the improvement.

If we build feedback loops across a global network we can get exponential speedups of technological development. What this means is that if I am a technology creator and share my creations freely, the thing I create can be instantly transformed into a thing co-created by a global community which is vastly superior to what I would have made alone. If my only goal in building technology is to directly convert the trash in my physical environment into useful things, my choices in regards to how I relate to the rest of the world will be based on trying to do that better and better. If the more I share the more this happens, often with what I make being totally replaced by something much better, my incentives undergo a radical shift. What I now want more than making a good thing and controlling it is making a thing which entices others to improve it. Again, as with money, we find that the idea of property inhibits us from doing what we need to do to build this new civilization.

The third leg of our tripod is mining. This is perhaps the most fundamental. The entire basis of our long strange trip from stone tools to bronze to iron to steel to silicon and so on is based on mining. We need mines to get the materials to make things. In order to make complex things with many materials, we need a global system which maintains physical control over all those mines using the system of property and the governments which uphold that property. We also need a global supply chain again based on stability of governments and large institutions to maintain the constant global flow of goods. In order for our current system to work, every individual element from oil to lithium to uranium must be transported to everyplace on Earth. Conversely, every piece of land with a resource on it will under the current system be pressured to extract that resource and push it out into the global economy.

Under the mining regime, every single type of mine is a choke point to the whole global system. This regime is inherently conservative: any threat to any part of it will make the whole system fail, harming everyone who relies on it which is presently everyone. In order to keep the system running, therefore, a constant global regime of military force is required. One cannot have a mining based civilization without military empires to control the large scale flow of materials.

And again as with money and property we have to examine how the system of mining affects our relations with our fellow people. As long as value all comes from a mine, we are all in competition at some level for the mines products. Every so-called "developing" nation will be forced by the dominant powers to extract all their resources to benefit someone else since all the nations are in competition to benefit from those resources.

But when we build everything in our civilization from the trash of the old world this situation completely changes. My motivation as a producer of technology is now to turn trash into things of value as much as possible. If you are thousands of miles away, and we never exchange any physical goods, just information, my incentive is now to have you replicate the trash technology as much as possible. This is for several reasons. First, there is the same reason as stated above in the discussion of property: the network effect. The more people copy my technology the better it will get, and the more comfortable my own life will become. But also, if we build a society which abolishes the mine, as long as other people are mining they will pose a threat to us. As long as anyone in the world bases their civilization on mining, they will need to build empires and dominate large masses of land in order to keep mining. So if I want to not be invaded by a mine-based empire it is in my best interest to help every other place on Earth develop the same technology in order to also prevent the violence of the mine from destroying what we build.

In today's world *every* single useful material we need for advanced technology has been pretty evenly distributed to every corner of the globe. This is totally unprecedented! There is nothing in history even remotely similar to the situation we find ourselves in today. People do not seem to have really grasped how fundamental and irreversible this is. Even if humanity died out tomorrow and were replaced by evolved crows in 100 million years, the distribution of rare minerals around the globe will remain. We will never again have to discover from scratch how to find and extract materials like cobalt or tantalum. We not only have all the materials needed to build an advanced technological civilization from scratch, those materials are already in a very organized form specifically designed to be useful. Aluminum has already been extracted from bauxite, iron has been smelted into steel, silicon purified into wafers of unimaginable perfection and so on.

Examining all three of these legs(money, property and mining) it should now be clear that these ways of thinking fall apart when we build all of our technology from trash instead of mined materials. It is my intent in this work to build a framework for creating this new world. To do this, we will need to build up a whole world of thought and action from scratch. The most fundamental focus of this new world is replication. We study how media replicates, how machines replicate, how software and hardware replicate, how whole systems replicate, and how pure information replicates.

We also take as an axiom that geometric thinking is more fundamental than numerical. This is because geometry is what we use to actually make things. From buildings to microchips to injection molded plastic enclosures, all technology is essentially a geometric construction of one kind or another. So if we are interested in building media the sole purpose of which is to replicate technology effectively, we find that geometry is the most fundamental form of mathematical thinking. As with the existing system of thought we currently live in, we will need to delve deeply into our most fundamental assumptions of how the world works. But now rather than trying to find some abstract truth, as the mathematicians of the early 20th century did, we build up a system of thought based on outcome: that which freely replicates useful things from trash is the goal, whatever that turns out to be. This will lead to re-evaluating how we think about machines and mathematical philosophy, replacing the theories of "computers" with ideas about geometric machines to print symbols.

One final note to make about this geometric world view. As with all our mathematics in this new civilization, our goal is replication, not finding some higher truth. This means that geometry is all based on its meaning to humans. Even if the meaning is in the angle of a turbine blade which communicates a different level of air movement in an air conditioner, it is this meaning we care about, not some abstract theorem to prove or algorithm to own. We therefore take language and symbols to be the most fundamental elements of which our Universe is constructed. We accept that whatever we may think or do, the "real" universe is separated from our minds by a veil of language we can never fully see clearly through. Reductionist science has made the mistake of ignoring this veil and focusing on a hypothetical "objective reality". Whether or not this is a permanent intellectual dead end is of no interest to us here. We want results, fast. We want a better civilization in our lifetimes. And to do that we build up a new way of thinking about information where our desire to provide direct value to people and replicating that to as many people as possible is our most fundamental axiom.

We call this geometric system of value, this geometric meta-langauge, Geometron. This is the Book of Geometron, which describes how to replicate the whole system.

To build this world we will first discuss how media needs to change to support this new way of thinking, then how we will physically deploy this media infrastructure to the streets of our world. We then show how the software works, how it replicates, and how you can add to it, improve it and make it your own to share. After that, we step through all the different layers of information that make up this new network. We then talk about our theory of being(ontology), the new underlying mathematical philosophy which we are using to replace the axiomatic set theory that 20th century mathematicians used to describe reality, as well as the theory of machines which replaces the Turing model of computation. Finally we use this idea of how information works to discuss the self-replicating set known as Trash Robot which we will replicate through the network and also use as a vehicle to help stimulate replication.

Chapter 2

Organic Media

In a consumption-based civilization the purpose of all media is to stimulate consumption. This can come in the form of advertising, corporate propaganda, or the legitimization of the imperial power required to keep mines and long distance supply chains operating. In the age of digital media we find that the hardware itself is also a large component of how it facilitates consumption, with planned obsolescence creating a stream from mine to landfill unprecedented in human history. A constant race to build ever more exotic materials and technologies into physical media devices creates a vast suction across the planet, forcing every corner of the globe to exploit anything that can be mined for digital media hardware from cobalt to lithium to be exploited as fast and widely as possible.

This state of affairs creates a powerful opportunity for a new type of media. The fact that the existing system pumps out this constant stream of new objects with all the stuff needed for advanced information technology creates a resource which can be used to build new hybrid technologies designed to incorporate scavenged parts from the discarded tech. The tech industry is now working on turning out *trillions* of "internet of things" devices–objects which have built-in networking capability and could in theory serve as web servers. Based on how these industries are structured we know that these will all be designed to fail on a very short time scale, and what they are selling today will be in a landfill in less than 5 years.

It is worth marveling at the scale of consumption built into the current digital media system before discussing the alternative. People walk around with screens in their pockets, the sole purpose of which is to manipulate them into consuming more. Those screens are built on a technology which uses the most exotic materials known to humanity, extracted at great human cost from every corner of the globe. And then they are forced into obsolescence within months in some cases by a software industry which is based around the idea of planned obsolescence. Having pushed their way into the pockets, homes and workplaces of something like half the humans on Earth, the industry is now pushing to put their devices in things which have no reason to be part of this network, like toasters and juice makers. The sheer insanity of this is hard to wrap ones head around, but because the entire media is controlled by this industry it is hard to even articulate in public how insane this is. And it is getting worse very quickly. The need to replace this parasitic monster with media which serves the needs of humanity could not be more urgent!

What we want from a media technology to build our new trash-based civilization on is to replace consumption with replication. We are now constantly buying new machines built from mined materials which constantly tell us to consume more things. Consumption-based media forms a consumption information loop. We want to form a replication-based loop, where the media is built from trash and contains the information required to replicate itself. We call this "organic media", because it behaves like a living thing. In fact it effectively *is* a living thing. If we built closed loop systems in which humanity is re-using material again and again forever in order to live in harmony with the world around us, it makes sense to think of us in combination with our media and the ecosystems we live in as a living system.

This idea of "organic" media is in direct contrast to "viral" media which dominates in social media today. In viral media, information replicates, just as viruses replicate themselves inside a living organism, but the this is always happening in a space defined by a fixed media entity. Social media platforms encourage information to replicate as fast as possible within their systems, as that costs them nothing and induces more people to keep coming back to their platform to be manipulated by their advertisers. But if someone tries to replicate the media platform itself by for instance trying to start a new platform, they will do anything they can to stop them. Media today can be viewed in biological terms as an apex predator which kills everything in the rest of the ecosystem and is full of viruses.

When we say we want media to be organic what we mean is that we want the media platform itself to replicate. Just as each new tree or squirrel in a forest really is a whole new instance of "tree" or "squirrel", with no central entity controlling them, we want each new instance of our system to be self-contained. We want it to be able to replicate itself right where it stands, with no outside input from some central system of any kind. This is only possible because of the waste of the existing industrial system: all the materials needed for advanced information technology are sitting in trash bins, dumpsters, closets, and landfills within walking distance of wherever you are reading this right now. All you need to build a whole new media ecosystem from scratch is information: the information required to gather the people and materials required to build it. If the system you build tells people how to do this, it can freely replicate across the whole world without any central infrastructure.

It is worth noting that building this is hard. Modern

digital media technology is designed by hostile engineers to be as hard as possible to fix, modify, or use for anything other than consuming advertising for a few months before it goes to the landfill. Those machines are built by vast teams of well funded groups with extremely specialized technical skills. It will take a concerted research effort to fully replace the existing information technology system with a free one built from trash. In a later chapter, I will discuss how we can do this by using a different system architecture in which the purpose of the whole system is displaying of a specific class of documents based on the software presented here. But for the time being, in order to launch our new media system, we will rely on existing off the shelf hardware which is still part of the consumer system but not the main commercial advertising-driven part.

This book is therefore doing two things in regards to launching this system. It is launching a new social media platform based on using the Raspberry Pi as a local web server used over local wifi networks, and it is laying the conceptual framework for building a whole new information technology system from the ground up on new principles. Just as technology people in the existing system refer to a "technology stack" we are building a whole new "stack" in the sense of a collection of technologies which are all related by a chain of increasing or decreasing abstraction or proximity to the user which work together to make our system work. In this work, we describe the whole stack. We launch fully functioning software and hardware for parts of it, and describe how we will build the other parts that require more work.

The most important part of the current project presented here is that it work for its purpose which will support the rest of the development. This means that the technology has to work to distribute new technology of all kinds built from trash which people need or want, which can be freely replicated. We use this partially consumption driven media system to launch self-replicating media systems which really are built from trash as a demonstration. Our metric of success will be how this self-replicating media technology replicates and evolves. If we can make it replicate by making things people want, and make it evolve by creating a strong incentive for people to improve it, the system will naturally evolve into the one we need, which no longer requires any input from the mine-based system to function anymore.

Just as a relatively small number of people a few hundred years ago sending each other letters built the basis of the current explosion of technology which led to the existing world order, we believe that a small number of people with new ideas today can build a much faster explosion of information which consumes the existing world of consumption and replaces it with locally closed loops of material in a single generation. We also believe that this is of the utmost importance to do as quickly as possible. The existing system is killing us. It is destroying the natural world, needs constant warfare to function, and is increasingly driving anyone outside the technocratic elite into extreme poverty.

So how do we actually build this "organic media"? We start by looking at living systems, both individual organisms and larger systems like forests. The most fundamental thing life does is replicate. This will probably get tedious for the reader, but replication is the thing this work will come back to with relentless repetition but that relentless repetition of replication is precisely what makes life work. A living system is a system of thing which all also replicate. Living systems replicate over all scales: forests replicate, but so does the RNA and DNA in each cell of each organism in the forest! We will also build our systems this way: many components make up systems from tiny scraps of code or single cutouts of cardboard up through whole vast industrial fabrication systems, and we want *all* of them replicating. Again this is in direct contrast to the existing system in which small parts like shared memes on a platform are supposed to replicate but the company itself is designed around non-replication.

Another property of life is that it is an independently evolving thing. Because organisms have an independent life, they can change in much more unpredictable ways than centrally controlled systems like a large corporation, government or non profit(this includes open source software projects with a central code base that all instances are copied from).

Finally, all life dies. In order for life to work we need the cycle of death to be natural. Just as fungi in a forest turn logs into soil we need the destruction of all things in our system to be natural. This is again in direct contrast to the existing system in which all media is built out of a company which is designed to grow forever and never die.

In order to build our platform then we want to write down a set of rules which will guide all of our work. These are nine rules:

Everything replicates. This is the most fundamental law. It is what makes life alive. And it is what makes media organic rather than viral or parasitic. This means that all our software contains code to replicate itself without any reference to a central code repository. The code on a server in a coffee shop can directly replicate to the laptop or phone of every person in that coffee shop with no connection to the rest of the Internet at all. All our hardware is built into some kind of media which describes its replication.

Everything evolves. All things can be edited by all users. To be in contact with a thing, be it a file or a physical machine is to have the power to alter that thing totally. There are no "users" or "engineers" in the sense used today. Some people will choose to edit things more than others but everyone *can* edit all things.

Everything dies. All things can be deleted or destroyed by all people. This is particularly important for files. Much of the power structure we are trying to destroy rests on information we are not allowed to destroy, from the intrusive and parasitic industry of buying and selling personal data through the constant advertising we are not allowed to turn off. Also, in order to be able to stop harmful information, we empower every single user with no exception to be able to delete every single piece of information they come into contact with, with no exceptions. This is less destructive than it sounds. Because our network is all physically local, no central bad actor can wipe out the whole network. If all networking is at the level of a wifi network, and they are all constantly being destroyed and rebuilt anyway, the cost of universal destructive power is outweighed by the benefit of people being able to stop bad information without any appeal to authority.

No property. As discussed in the previous chapter, the idea of property is not compatible with a civilization based on self-replication. Since the idea of property fails at scale in a replication society, and since our goal is to scale, we dispense with it immediately and build infrastructure which not only has no intellectual property but where the physical web servers are not owned by anyone. Initially we will have to spend money to buy parts to build them, but as soon as they are built we will release them to whoever we think will get the most use from them, along with instructions for them to do the same, passing all infrastructure along to wherever it gets the most use. Building network infrastructure without property for the benefit of our communities means that our incentives are now to find whoever has the most need, identifying their needs, and using our technology to serve those needs and fast and directly as possible. If people benefit from the systems, they will naturally be able to replicate, which will further replicate the non-property technology. Initially this means building Raspberry Pi based web servers and giving them away to the people with the greatest need.

No money. This is connected to the rule against property. Initially of course we will need to spend money to buy parts, and will need to have users make money to support their near-term survival. But as we scale up and get more and more basic needs satisfied by technology built from trash, we want to have the elimination of money be the direction we are headed from the start. This is not as far fetched as it sounds. As will be discussed in the next section, barter will be an incredibly powerful tool for scaling. Our network can provide huge benefit to very powerful and wealthy people, and if our people with the most need are dispensing this benefit, we will be able to barter the things we need to scale directly. When our network helps a business with a lot of unused space to make money, they can let us use their space. When a business person makes connections using our network which make them money, letting us scale our software up on some of their servers will make economic sense to them. We will be able to barter our value as network builders into the things we need for personal survival like places to sleep and food but also the things we need to scale our technology like access to labs and machines.

No mining. Our long term goal is the global elimination of all mining. This includes the whole natural resource extraction industry such as oil and gas as well. This cannot happen overnight, but we don't need it to. Every single mined component we replace with one from a dumpster or landfill takes a little bit of energy and power out of the mining system. if we can remove power from them in a way which self-replicates, our system will simply consume theirs, and mining will be eliminated in a generation.

Everything is physical. This is almost a circular statement. What does it mean for a thing to be "not physical"? This is a statement of belief. We emphbelieve that the idea of information which is not physical is meaningless. All information has a physical manifestation, be it charge on a transistor or bumps on a CD. This law is important as a vocal rejection of any theory of how machines works which states that information or data can exist independently of its physical existence.

Everything is recursive. One of the most notable properties of life is its constant self-referencing. Billions of DNA strands in each individual body of a large organism all contain a whole copy of the information required to replicate the organism. We see information which points to information which points to information. Life is very self-referential and involves in an abstract sense functions which call themselves constantly. RNA stores instructions to make molecules which replicate RNA, and so on. This law is to remind us as creators of technology to be *constantly* thinking of ways to make things point back to themselves to replicate.

Everything is fractal. This is another property of living systems that we take for granted but which we either ignore or make very crude imitations of presently. Centralized systems of control create technologies which are flat in scale: we build microchips with nanometer precision across millions of nanometers(mm) of scale and so on. In contrast, living systems are fractal in scale such the scale of "error" required to cause catastrophic failure scales with the size of the system. We assume that patterns will repeat again and again at different scales, and expect that our technologies only need to be precise at the correct scale for any given sub-system. This has very specific implications for fabrication which will be explored elsewhere, but as a law we mean that we must simply always have ideas of the fractal nature of living systems in our minds as we create new things in our new trash-based civilization.

With our goals and laws of operation stated, we are now finally ready to delve into more detail into what we are actually constructing with this work. This is a local network based on a web server loaded on a Raspberry Pi. The Raspberry Pi is a computer on a circuit board about the size of a deck of cards which typically costs about $50. It needs some peripherals including a screen, keyboard, mouse, battery, and memory card, all of which makes it about $200 for a nice, easy-to-use, portable and self-contained system. It only requires a few commands which are easily copy/pasted to install the whole functioning Geometron system on a new Raspberry Pi. It is also modular, and if portability is not needed it can plug into the wall, borrow a keyboard from another system temporarily, and display on a TV, making a non-portable system cost just the value of the board($50).

It is also important to note that this is all modular, easy to buy from many sources, and involves parts many people already have lying around. The Raspberry Pi is widely marketed as a hobby tool and a STEM education tool, but its use case is not always clear. Therefore a large number of people own them but do not use them. They are often sitting in drawers in someones home office or a under-used maker space gathering dust. If we have a use for them and can provide useful services to people it should be possible to directly barter with people who want to support our network who will be willing to donate the Pi boards to our network, where they will become non-property and be distributed to those in need.

The actual software we run on the Pi is what is described in the bulk of this book. It is all designed to run in a web browser. Any web browser. So if a Raspberry Pi running the Geometron server software is on a wifi network, all the programs and documents on it can be read, used, edited, replicated and deleted by anyone connecting to that wifi network on any device be it a phone, laptop, tablet, or another Raspberry Pi. Also, the pi itself can serve be used in the same way as any other device on the network.

We must also note an important condition for the pi to be a non-property computer. In order for the pi to be able to freely be shared among the people, it cannot have any personal data on it which causes someone loss if it is read by another person. That means we must never log into private networks like gmail, facebook, or more sensitive things like bank accounts ever on the system. In order for a free and open system without property to function, it must be kept separate from the property based networks. This media platform exists for the sole purpose of free sharing of documents. Any document we do not wish to share we do not put on it.

Also, there are no "users" on this network. This is a network of documents, not users. There are no logins, no passwords, and no databases. User data is not harvested for profit because we simply do not generate the type of information which is considered "user data" in the existing systems.

We take as an axiom in the development of this system that documents intended for free sharing are of greater value than private documents, and that the network effect as the universe of non-property documents grows will exponentially increase their value to people until our network out-replicates the existing ones.

Our network is based on sharing several specific types of document which are encoded into the software. We share "scrolls" which are text documents, "maps" which are like presentation slides or memes, "feeds" which are essentially lists of information, self-replicating applications of all kinds(using web based code that runs in a browser, never native code), and generalized Symbols using the Geometron geometric programming language which takes up much of this book. Taken together, these systems of document creation and replication will allow us to describe and replicate any technology of any kind.

This network will be distributed physically, over the Street Network described in the next section. We will initially try to get servers to people who are on the move, living on the streets or in vans and buses, truck drivers, street performers–anyone who finds themselves in nodes of physical networking like dense urban areas or truck stops with many people passing through. The details of how to replicate the server are in the section after next, which delves into the code. Not everyone will need to read this, but everyone should know where to find it, as part of what we will be bartering for as we scale is help from technical experts who can learn the system and replicate the software as well as add to it to evolve it and decentralize the code base.

The rest of this book describes how to replicate the system, how to use it, how it works, how to develop it into a fully trash-based system, and how we will rebuild mathematics to support this venture.

The Book of Geometron

This book itself is organic media. It is intended to teach its contents to a reader(or rather a small subset of readers) to the level where they can then teach another. This should enable them to re-write future improved versions. In this section I describe how the book was put together, where the files are stored, how to edit them and use the LaTeXdocument preparation system to make the files required to produce a finished book. This means you also need to know what is required to get the physical book printed at an on demand printer, get all the metadata required for publication and distribution, and sell your version in retailers both large and small and online and off. Thus even the book is fully decentralized in principle: if it costs you nothing to set it up to sell, you can sell only a half dozen copies and it will be a net positive, and then if the next person does this it will also be positive and so on.

If the book is decentralized in this way of distribution it has many advantages. If the book turns out to be disruptive enough that people try to use lawsuits to shut it down or harass an author, but there are 10's of thousands of new authors popping up all the time, it will be impossible to shut down. As some versions turn out to be dangerous or illegal, other versions can immediately be published with omit the offending content. Also, many editions will mean some will get much better than this initial manuscript. Decentralization means that as the manuscript finds its way into communities that speak different languages, the translation can happen without any centralized effort. So for instance if someone translates from the original English into say French, and then it spreads around in areas bilingual with French and some other language like Swahili it can go directly from the French to the Swahili without any involvement of the initial English speaking writers. By avoiding copyright, these improved and translated versions can then get translated back into English and sold yet again under yet anotehr edition. Having editions be unique can create a market for unusual editions, further pumping money into the system and stimulating further development of the book. I would rather see 10,000 people make 100 dollars each selling their own editions of this book to just their friends than see me as the initial author make 1 million dollars on 500,000 copies of one edition.

It is not my intent in the long run to make money on Geometron. It is my intent to create a network which allows us to live without money by directly bartering what we need to survive(food, a place to sleep and work, medicine, transport) without use of money or any production in the old consumer economy.

All editions are published with a public domain license for everything but the final pdf. The final pdf is published under the minimal copyright required for an author to create the needed publication metadata to get distribution outside of the on demand press used for printing.

Each chapter is a .tex file, using standard LaTeX.

This work must replicate itself completely. We show here how to edit each chapter, publish them to a public Github repository with detailed instructions for further replication, compile the document to a .pdf in book format, and self-publish the fully compiled book on Lulu Press. We then guide the reader to follow the instructions on Lulu to get all the needed copyright metadata for official distribution through normal publication channels. We then describe how to order just a few copies, sell them along with other parts of the system here at a markup, and use the profits to buy more print copies of their own book to place in bookstores and libraries as a fully guerilla activity with no official sanction. This is a little twist on the methodology of Abbie Hoffman's "Steal This Book." In Steel This Book, book sellers had to buy the book, which readers inevitably stole. cutting into book store profits. We use guerilla production methods to distribute it into bookstores without them spending money. They are faced with a choice: go along with our program and take free money from customers for the book or make trouble for us, throw the book out, and loose what is for them totally free money. If they sell the books for a profit, it benefits our network, because it spreads our ideas but also because it creates a value stream in the existing economy based on what we create, which gives us power in that network. Bookstores want money. But we want network centrality and the ability to control how information flows in a network, and free distribution shifts the power to us. This is why the self-replication of the book is so important. If you want to make your own spin on this book and make it more of a best seller, you do it, but if you leave it open, you hope someone else does it again, and that it keeps getting better as it replicates. Replication instructions for the physical book will live in the README file for the book repository on Github.

Another way to make the book self-replicating is to distribute print copies which a number of blank pages, sold at cost. Artists can buy these, then create illuminated manuscripts with custom illustrations of geometric art using Geoemetron. This can then be bartered based on the needs and powers of the artist, with outcomes as diverse as the artists themselves. We can use this to barter for what we need to expand the network, to survive, and just as another way for artists to exist and thrive in the world. Also, this can be passed along and shared as non-property. You can buy the hardcopy of the book, add a little art, and pass it along, then the next person adds some and passes it along, and we all help each other out along the way. A famous artist or tech creator might be able to barter for something of very high value. Each physical work will be unique. And since the electronic files which generate the pdf which goes to the hard copy also self-replicate, this entire system can evolve and replicate, as dedicated creators rewrite the whole manuscript and add their version on Lulu press for sale at cost print-on-demand as well, and so on.

Chapter 3

Street Network

What is the Street Network?

The consumer society we live in today relies on a vast and well-organized collection of networks to function. There are the networks which channel freight from one place to another, like trains, trucks and boats. There are the human transport networks like planes, personal cars and buses. There are the invisible networks of power based on who when to school with whom, family connections, or business connections. And of course there is the global information network centered on the Internet which connects everyone together by information.

What we seek to build is the means of replication of trash-based technology in order to propagate our new civilization built entirely from trash. This new system will be much more localized than the existing one. Rather than needing a constant high rate of movement of very large quantities of physical goods, we plan to build systems which will ultimately use material directly available in our physical environment rather than from far away. Even in places with limited agricultural capacity, I believe that with the superior technology which we can build using a more organic system that we can build dense food production and water purification everywhere once we put our minds to it, making almost all large scale movement of goods un-needed. We will still move goods and people, but more by choice than necessity for personal reasons. When globalization does not force everyplace to be identical, travel can become an adventure again!

Part of the ideology of the existing Internet-based culture is that place no longer matters. People appear on a teleconference or send an email and no one cares where they are. People arrive for a meeting in a business hotel or conference center and it is identical to every other place in the world. The ideological basis of the Geometron Street Network is that place *does* matter and *should* matter. We reject the idea of "nowhere". Everyone is somewhere. Whether you are in a rural area, a suburb, a big city or a highway rest stop you are always *somewhere*. Furthermore, that place you are has its own local geographical logic. It has its own crossroads, its own nodes of power and connection, whether it is the local pub or coffee shop or the exit of a subway station. All these little places on the scale of one human body have real meaning for the people who inhabit that space.

We want an information network based around physical replication of technology from trash. To stimulate the replication of the Network, we need it to create value for people who use it and operate it. This value can be of many kinds: it can directly provide physical goods people need, it can facilitate business in the monetary economy, it can provide mutual aid to a community, it can create local social connections, can build network power for users, and any of these values can be traded for materials and space needed to continue to expand the network.

The Street Network consists of the people going out and spreading all this, the web servers we use to do it, and globally visible web pages which serve as links to connect users to us and our network.

We buy domain names which are linked to a place but not property. We avoid .com and focus on .org, .net or .xyz domains. We avoid specific addresses or names of companies. We choose names that describe shared resources which are public enough that no one is in the position to claim ownership of the name. This can be the name of a neighborhood or street, a body of water, a park, or a mobile shared resource like a mutual aid bus used by people who already do not use property. In the physical location described by the domain we create and share physical media which points to the domain. In its simplest form this can just be a hand written cardboard sign. However, we can also use Geometron to make various self-replicating physical media which transmit the domain, such as laser cut spray stencils and the self-replicating clay tokens described later in this work. We also can fly a black flag of the place with cut out colored felt sewn on a black square of cloth, which is described in the Scrolls distributed with the system.

Each domain is hosted on some commercial server for the time being, from providers such as Dreamhost or Bluehost. Free web hosting can also be used at 000webhost.com and is recommended if you have no money. The Geometron software can then be installed on each of these servers according to the instructions in the next section(and repeated on each Geoemtron instance in the Scrolls), making it just another identical instance of the software to what exists on all the local Raspberry Pi based servers. The primary purpose of these globally visible domains is to point back to the local server. In its simplest form this is just a description of when and where you might find the server. Photographs of the network Operator and their gear can validate the system to passerby. The Map format described in a later chapter can also be used to precisely identify physical places where infrastructure can be found.

The deceptively simple structure described above is the Street Network. We are adding digital media technology to the oldest network in the world: the physical paths of movement. We will use this to follow all those paths, from superhighways to ancient footpaths to natural harbors, just as other ideas have traveled throughout human history. If we find the most powerful nodes of geography we can build a network of staggering power with a relatively small number of people initially.

Users

The uses of this network are very different for different people, just as is the case for the existing global networks like the Internet. In this section I will discuss some of the different user groups and how our network can provide value for them.

Operators. We are the start of this. Get a Raspberry Pi and install the system. Get a domain, install the system and point to your server. Go forth and share! Ultimately those of us who build and share this system will form a very powerful network of mendicants. The mendicant tradition has appeared many times in many places in history. A mendicant is someone who is totally devoted to their faith(they are generally religious orders) who renounces wealth and travels with no possessions asking passerby for donations to support them. This has traditionally created contradictions, as these orders have a way of gaining power and becoming anything but poor as they scale up. As our consumer society has destroyed itself it has driven more and more people into this way of living against their will. If our network provides vast amounts of value to people we will find that the most marginalized people of today when leveraging the power of this new network can barter for not just survival but to thrive in a new civilization without money, mining or property.

We follow the laws of Geoemtron listed in the previous chapter as a guide for building this new world. We teach everyone we meet how this whole system works, and recruit new people as Operators. Note that the idea of a mendicant order has strong religious overtones, but that this is a completely ecumenical order based on the universal language of geometry. Geometry has a central place in all existing cosmologies, both ones considered religious and ones considered secular. The work here presents a way of interacting with the world based on geometry. In some ways this whole project can be thought of as the start of a free school for teaching a new kind of geometry. This is a distant descendant then of the geometry schools of the ancient world. We teach people the whole system; mathematical philosophy, robotics, code, all kinds of industrial fabrication, crafts, fashion, whatever we build we teach and share freely.

Do not misinterpret this idea of the mendicant as a vow of poverty. We will be more wealthy than anyone currently living in the consumer society once we scale this Network. We are building a new world in which no one is poor. By starting from a baseline of people who have nothing but building better technology than what is presently available in consumer civilization we start by making sure those who have the least have everything: free clean water, free good food, free high technology medicine, free transport, free shelter, free network technology, free air conditioning and heat etc. If those who have the least have better stuff than the richest people in today's world the world of today will dissolve and be naturally replaced by this new order built from the waste of the old.

Traveling kids, hobos, panhandlers, people asking for money or selling things on the street corner. A physically local free bulletin board shared by passerby in a high traffic area can allow people asking for money who are currently ignored by passerby as just another anonymous face and cardboard sign a chance to really tell their stories and to share all that they have to share. When people share their stories they can become part of the emergent physical community of passerby in a location where the network node is located. When people view others as part of their community they not only are more willing to help, they can have open communication about the best way to help, expanding from just spare change to more comprehensive mutual aid. Because we clone content from the local terminal to web pages on globally visible domains linked to a physical place, which are advertised everywhere in that place, marginalized people whose only ability to get online is the public library can use the computers there to get the information they need to better survive, and ultimately to thrive and build new communities where they already are. The way a local network can help people is twofold. First, it is direct, by asking for money and other mutual aid. But by being physically on location all the time, already with physical media(cardboard signs), people in a given place can aid the network, creating value for the other people in the community who are more resourced, who then no longer view monetary support as "donation", but rather as an expense which supports their other business activities.

In order to see the power of this second means of network support of marginalized people on the street, we have to look more closely at the network nodes we are building. One of the major types of node is in a business district of a city where there are both homeless people asking for money, on the street all day with physical media, and power brokers who make their living entirely from connections. These people include venture capitalists, entrepreneurs, lobbyists, consultants, and the rest of what might be called the "deal-making class". An example of this confluence is some of the parks along K-Street in Washington DC. K Street and adjacent streets is home to a huge homeless population as well as power brokers whose livelihood depends entirely on connections. If a physical network were built which facilitated direct communication between people along K Street, the people who spend the most time physically on the street can be brokers of information on a network which can be worth a lot to the people who trade in information. Physically local information networks can leverage the power of physical places with very powerful people walking past all the time who normally never communicate. Connecting these people up can be dangerous. But if we provide them with value, it can be worth both a lot of money to them and also potentially something they can barter for giving us space to live and work nearby. If you facilitate a 10 million dollar deal and the customer knows you can do it again, the least they can do is give you a 100 dollar gift card to the nicest restaurant in the block. There is no real upper limit on what an enterprising Network Operator could in theory make if they learned to really channel information efficiently in the nodes of global power. And of course we must remember that when dealing with power brokers their currency is not money. When the people who currently have the most power in society find themselves dependent on free open networks, those networks themselves will gain power which penetrates that of the existing power structures, potentially creating an existential threat to them. We must take note of this.

The elements of traveler culture which overlap with "van life" are also key to increasing the network effects of the Street Network. This also links to trucker networks. People who live their lives on the road can use this network infrastructure to set up complex networks and markets in highway rest stops, Walmart parking lots etc. using either wifi networks in these places or their own hotspots from their phones. These networks can be of utility to passerby of all kinds, from tourists to truckers to the workers who keep the places running. Just as existing global social media networks provide value they can charge money for, a physically local network can provide value which people will pay for. An example use case here is a Street Network Operator agreeing to maintain a backup of and keep posting an advertisement for something a local entrepreneur is trying to sell to truckers. In exchange for that, they can get directly compensated in gas, right there in the rest stop, without money changing hands.

Food not bombs, street outreach, harm reduction people, mutual aid workers. See above. The people who are working to help the most marginalized members of any given community can better reach that community if there is a physically local media platform where people can share information about resources. Documents can be posted which explain how to get access to resources, when and where resources will be available, etc. Because the whole system self-replicates, as with Food Not Bombs, anything which is successful in any given place can be immediately cloned to other nodes on the network. Food Not Bombs already has a global network of free and open nodes with no property but a very recognizable brand identity and set of behaviors and actions. FNB nodes are generally already linked by networks both online and via people who travel from one punk house or FNB house to the next. The whole anarchist network of community houses, FNB's, anarchist infoshops and bookstores, really really free markets, free boxes, etc. can form a basis for a truly free information network carried from house to house and city to city, running on house wifi networks.

Business owners in a shopping center. Every business owner has neighbors who are also business owners. You already have an informal network. But installing free digital media infrastructure can provide huge value by allowing more mutual aid between neighbors of all kinds, both owners and customers. Tech giants ignore you. They demand monetary tribute in order to even have your business listed, and then still refuse to give you an equal footing to the corporate giants which dominate their platforms. By controlling a local platform in *your* shopping center, you can provide value to customers with articles they write and share with one another which brings them in(just as advertising-supported media has interesting content to get people to look at ads). And then this medium can have much more than just the ads you would get from a Big Tech platform. You can post really detailed information about everything you do with no restrictions on length, and share across the shopping center. If you own a karate school next to a dentist, the bored people in the waiting room can read about the history of karate right next to a detailed schedule of your class offerings. And when parents wait for their kids to get out of karate class they can be reading about clean gums in an article written by the dentist. Big Tech doesn't care about you. If you build your own network, you can center it right where it belongs: on the people actually using it, rather than a few oligarchs in San Francisco.

Coffee shop owners. Building a network in a coffee shop on the wifi network which requires purchase to use and which has a time limit can create a huge amount of added business for any local business owner. It also builds community. So coffee shop owners who find themselves with a full shop of laptop drones with headphones on who work for hours, or get kicked out and do the same thing somewhere else can instead find themselves the brokers in a very powerful information network. Much of the commerce of the world is now code written in coffee shops on laptops. Creating physically local networks around these already existing groups can create huge power for the users which then benefits the people who set up the infrastructure(again, just like existing centralized social media platforms.)

Web developers. We need web developers(people who can write HTML and JavaScript code) to be constantly writing more and better software in order to make Geometron a success. Developers who work all day in coffee shops or any other shared space like a co-working space or pub can have a social network based on both co-developing applications useful to all and sharing other resources. Developers will use the resource of the Street Network terminal/server on the local network in the same basic way as others: they can share their resumes, links to pages of personal projects. Developers are key to the whole system. We must recruit developers with this book who will rewrite all the code and also the book, replicating the whole system. The faster our network can get developers into the swarm, the faster the code itself will improve. Developers are key!! Developers create servers to share into the network. I now ask the reader to look up "steve balmer developers" on YouTube.

Power brokers. Venture capitalists, financiers, entrepreneurs, deal-makers of all kinds, lobbyists, politicians. Your network is your power. Geography matters. Build a network in the lobby. Post things on street nodes, build your network, build your power, build your literal street cred. Deal flow. Deal flow the likes of which you have never seen. Leverage the power of the physical street!

Crafters, makers, jewelers, artists. An alternative to Etsy, street vending, or being in a shop. Post your stuff to the local networks. This is much more free and long form than existing platforms, you can post images, descriptions, contact info, times and places when you'll be in a place. This can be way easier than other sales channels for arts and crafts. You can say when and where you'll be at a place, post a link for contact, and then show up in the network node like a coffee shop to make the physical exchange. In many cases, because the network is physical and local, there will be barter opportunities as well as direct sales. A barter economy can develop where people donate materials you use for your crafts as part of how they pay for the finished product. Removing shipping or transport costs by dealing directly in a physical location removes friction from the market, amplifying dramatically the power of the market, especially for crafts which involve physically bulky objects. For instance, people can bring in motors and properly prepared plastic sheets and cardboard, as well as rolls and rolls of duct tape, and we can exchange finished products built from these materials and tools, as well as free food, drinks, and supplies, creating a market economy without money as well as without formal business structures(making it easier for marginalized people to participate).

Any labor pool of gig economy workers focused on a specific geographic location. One of the most obvious of these is the drivers who presently drive for the major rideshare apps who all congregate at the airport to pick passengers up in the same exact place, and yet all of it is currently coordinated via the apps(unless you do the cab line). The rideshares apps have proven that cities will ignore illegal cabs if they're done at scale. It would be straightforward for a small team of Network Operators to run a server which replicates to a page which is advertised around, something like a domain of yourairportnamerides.xyz, which tells users how to log onto the wifi network created by an Operator's hotspot near the pickup zone and with a link on the page to the local network address of the server. All all this IT is doing is directing customers to a dispatcher who manages the drivers over a simple app shared by the collective. The whole network is run by a team of about 2-4 people. One person might be a developer, who creates the app to manage all the drivers and post messages from dispatch. Another person is all marketing, putting up the relevant information in the right places to get seen by travelers but not stopped by the rideshare apps, airport authorities, or the cab companies. Riders will never have their destination information on the public network, nor will drivers put personal information, but they can work on an open trust model where they are known by dispatch, who has code names for them, and operates a queue app which simply adds drivers as the arrive near the Airport and pushes the most senior driver to the top of the stack, which is passed along to a rider. Another Operator might be the one who runs the trust network for the drivers, verifying everyone and organizing meetings for the whole cooperative. This can be used to unionize existing workforces quickly as well, building ad hoc networks which are very hard to suppress visible to everyone on their mobile devices on a local wifi network.

The same model holds for places where workers congregate looking for short term construction work. Those locations can have a server where an Operator runs a labor marketplace where a much larger and deeper labor pool can now advertise, but without all having to be in the physical location. This means a crowd of a dozen workers looking for work can be replaced by an Operator with a sign pointing to the domain where the copy of the market is hosted. Workers who come by can leave an ad on the local Raspberry Pi Geometron server, and anyone coming by looking for construction labor can just scroll through a now much deeper collection of ads and call whoever they need to hire. A market place like this can suddenly go from a dozen general laborers to a construction labor market which includes specialists like plumbers and electricians as well as much larger general contractors just looking to save on marketing costs. A person holding a cardboard sign on a street corner by a giant box home improvement store can now potentially be the broker of an information network on which millions of dollars of commerce flow.

Trash Robot

The Trash Robot is a self-replicating set of things described later in this work. It consists of an open brand combined with a way to create self-replicating symbols which represent in principle anything one can express with language. We use the Geometron system and the Street Network as a vehicle to distribute Trash Robot.

Trash Robot icon printers form the basis of a symbolic economy. This means that we construct self-replicating physical media which can have symbols representing anything, and we use these symbols to communicate with each other about how to replicate things. In a numerical economy we exchange money for goods or services based on a numerical evaluation of the "value" of those goods or services. In Trash Robot we use the Geometron language to make constructions of pure geometry which can be used to organize our thoughts and discussions as we share with one another self-replicating technology of all kinds.

Trash robot consists of a collection of methods for building the robots and operating existing machines to make all these icons symbols, as well as the open brand of the fashion and accessories and arts and crafts which symbolize the system. This brand consists of googly eyes, rainbows, black cotton flannel, cut felt and rainbow duct tape over cardboard and bamboo. By being a very recognizable brand identity which is generic enough to be impossible to copyright, we make a vehicle for technology which is not property to freely replicate.

Chapter 4

Servers

Servers are machines that store and share documents in the Geometron system. There are three kinds of servers we work with: the Raspberry Pi servers that form the real backbone of the network, globally visible hosted domains used to point people to local Pi servers, and developer servers for editing and sharing new versions of the Geometron software itself.

To spread the Network, the most fundamental form of replication is replicating the local Raspberry Pi server. To do this, we want to either buy or barter for the parts, install the software, and teach someone how to operate it so that it can be sent out onto the Street for public use. Given the choice, we will always barter for the parts. If we can find people who support our cause and have extra technology hardware like old keyboards, screens, or even Raspberry Pi boards, it will be easiest to take direct donation of hardware on location by an existing server than to deal with purchasing and shipping new hardware.

The elements of the basic Raspberry Pi server are:

- The Raspberry Pi board itself. This is a circuit board about the size of a deck of cards. All of them should work! Older ones might be slower but they should all work. This has all been tested on the Pi 3 and Pi 4. Boards are generally between 40 and 50 dollars, but again if you can barter them that's ideal, as they are often sitting idle in peoples desk drawers.

- micro SD card and SD card reader to write the card.

- USB keyboard. If possible, find the one without the number pad so that it fits more easily in a backpack, this can make a huge difference in portability.

- USB mouse.

- HDMI Display. Not all work, but all can be made to work. Ideally you want a small screen which can run off the same battery as the Pi. If you are setting up in a fixed location you can use any standard modern TV screen, which good both for needing fewer new resources and for visibility. A large screen display for a server can be a great way to have shared physical social media, where people can read documents on the big screen without needing any device of their own. For small portable ones we recommend buying ones specifically sold for the Pi, and which say that they don't need any special installation of software to work.

- HDMI or HDMI mini cable. There are some tiny Pi displays which don't need this because they connect to the pins on the board. Note also that you will need the HDMI mini cable for the Pi model 4 but the regular HDMI for all other models.

- Lithium ion polymer battery packs. This is only for portable installations, but it is important to have something modular and portable with a lot of power storage capacity. We recommend the TalentCell batteries as being easy to charge, easy to carry, and having both 12 V and USB output. They also sell solar chargers for those batteries, which are useful to have for long stretches of being away from power.

- Wall power. For the Pi model 3 and below this means a wall supply with the same USB micro and for the model 4 it is USB C. You will want a wall supply that can put out 3 amps at 5 volts. Also, if you are running everything of 12 volts you will already have a wall supply that came with the battery packs listed above.

- Wifi hotspot. This can just be a phone with the hotspot turned on. But it can also be a mobile hotspot from the phone company which has its own wifi and connects to the network.

When you have all the materials to make a server, you will want to start by setting up the Raspberry Pi in the normal way documented on the Raspberry Pi website. Follow the instructions on there to copy the NOOBS operating system onto the SD card. You can also buy SD cards with NOOBS already installed. The Raspberry Pi home page is www.raspberrypi.org. Buying Raspberry Pi stuff can be done in person at some electronics retailers, some maker spaces, and online from numerous retailers, just search and look around. Sunfounder is a great source of compact portable screens for the Pi as well as other Pi things.

Note that part of setting up the Pi is logging onto some kind of wifi network, which is similar to on any other computer system, you click on the wifi icon in the upper right of the screen, select a network and put in the key. If we are using a hot spot we will want to select a simple name and password like using "geometron" for both, and post as widely as possible to potential users what that is so it's easy for them to remember and log on. In order for this network to be visible, all users must share a wifi network. It is also possible to point a global domain name to a local Pi visibible to the outside world, but that is beyond the scope of this book.

Once the basic operating system is set up(set it up with no password) you will want to install the web server, the PHP language, and the Geometron server. To do this you can follow the instructions at github.com/lafelabs/thing, which is where all the code for this project lives and from which it will all be copied when you install. This copying process copies all the detailed instructions to copy the system, so if you find any instance of the Geometron system on a global or local server you can follow the instructions on there to replicate.

When the Geometron server is installed on the Pi you can interact with it by opening the web browser on the Pi and pointing it to http://localhost. This should now look like any other server in Geometron. This can be used to create, edit and replicate documents of all the formats in the Geometron system, which are documented in the next several chapters of this book, as well as on each instance of the system.

When you set up a new Pi server, you will want to copy the Pi scroll to the home scroll, and there is a link to do that in the default screen. When the Pi has a correct Home Scroll there will be a link to open a page which makes a QR code for the server. You can then scan the QR code on the screen to log on with any mobile device which is logged onto the same wifi network as the Pi.

The second type of Geometron server is on remote hosted domains. As discussed in earlier chapters, we will buy domain names based on generic places that are not owned by anyone but have a physicality of some kind. For example streets, parks, rivers, neighborhoods, truck stops, or mobile mutual aid stations. To install the Geometron server on a globally hosted domain, just copy the file replicator.php from any existing Geometron server then point a browser to it. This is documented in the README documents.

Finally the developer server is used for local editing of Geometron on a private computer which can then be pushed to public repositories in a host like Github. This is done using the built in web server of the PHP language. If you are using a Mac, PHP is built in and you can run all this from the command line. If you are on a PC you will need to install the Ubuntu machine under Windows 10, install PHP and use that command line. In either case, start a new Github repository, set up whatever you would normally use for development via Github and put the file replicator.php in there. Run it with php replicator.php at the command line. Then while in that directory, run php -S localhost:80 and point a browser on that machine to http://localhost. Now you can operate Geometron as normal.

To edit all the code on the system, use editor.php, which is linked from the README file. To point the next replicator to your new instance of Geometron, edit the code in php/replicator.txt to point to the /data/dna.txt file of your new instance, and then convert that to php with text2php.php, which is linked from the code editor.

The details of the system should be documented on the system itself and on other linked media, so it is redundant and tedious to delve too deeply into technical details here. This is just here for completeness so that the system is described in broad strokes and so there are pointers to all the bits you need to learn about if you want to move from just using the system to building your own new systems based on it.

These three types of server are the whole system! There is no company selling server space or running a central code base. Each individual server of any kind has the whole system on it. If every system on the planet were deleted in an instant, you could repopulate the entire world with the one copy on the individual Raspberry Pi you are using, or the Github repository you cloned it from or the web page of the local street that connected you to the Network. Github repositories replicate code, hosted domains point to Pi servers, and Pi servers are the medium on which all documents can be created, replicated, and shared freely across the whole rest of the Network. All this can happen with no company, no organization, no centralized code base, no centralized brand or naming convention, no authority, no property, no cash flow, and no presence in any app store. And all of it is open, clear, self-documenting, and easy to copy.

Now, go forth and multiply! Let us first make a million of these with the Raspberry Pi, then learn to make them on old hardware which we install stripped down Linux systems on, and then finally on Geometron hardware built entirely from trash using the full stack Geometron system described later in this book. Millions of servers can serve hundreds of millions of people When we scale to using all reclaimed trash for hardware, we can scale to billions of servers, eliminating the personal layer of networking completely, with ubiquitous open and free media shared in physically public spaces. This technology is clearly already possible to build if we choose to do so, and it can clearly be done for free given the very high rate at which the existing consumption-based system is pumping out electronic trash with all the elements of media(screens, batteries, radio transmitters etc).

Chapter 5

Scrolls

Scrolls are the text documents of Geometron. Think of this them as the Microsoft Word of the Geometron ecosystem. While most people are of course familiar with text documents in formats like Word, the form of Scrolls and how they work in our system is quite different and requires a fair amount of explanation.

One of the most important requirements for a system of self-replicating documents to function is that documents be able to be copied to a clipboard from one window and then pasted into another window from the clip board without losing any information. This is not possible in formats like Word, since the system hides all the formatting from you. In Geoemtron, the Scroll format uses raw text with a few additions that allow for things like images, links, headers, and lists to be called using simple syntax. This is what is called a "markup language", and is the same class of languages as HTML, the "HyperText Markup Language", which is the basis for how all Web documents are formatted.

What we use in Geometron is Markdown, a markup language specifically designed to be as "lightweight" as possible, replacing the cumbersome tags of HTML with simpler and shorter syntax. Markdown was created by Aaron Swartz and John Gruber in 2002-2004 precisely to solve the problem we need solved: the simplest possible way to create a rich text document from plain text. One of the important differences between Markdown and HTML, however, is that if you start with a blank document and just type some text, not knowing anything about Markdown, that will be formatted and readable, without adding any special code at all at the start and end of the document. Markdown is also useful because it is used in several other open technology systems that are useful for us to work with in Geometron, namely the documentation on Github and the text cells of the Jupyter notebook system popular in open source science.

In addition to basic typesetting capability, we use the LaTeX typesetting system as an optional add-on for creating and sharing technical documents which use math. LaTeX is also used to build more complex documents which are in other formats, such as this book, which was written entirely using that system to convert to a .pdf format for printing. This technical document sharing capability is important if we wish to have our network form the basis of a new technological civilization: we will need some very technical documents to make that work. Also, LaTeXis used to typeset the equations in Jupyter notebooks mentioned above, so the ability to work with it is important for integration into existing open science projects.

Since everything in Geometron takes place in a Web Browser, we need a way to translate this raw code that we interact with when we edit and share files and something that looks the way things should look for people to read it. We do this using two open source JavaScript libraries which can be called remotely from a browser. To convert between Markdown and HTML we use Showdown.js, which is published under an MIT license, is easy for programmers to use with simple copy and pasting of code, and widely available. To convert LaTeXcode into HTML we use the the MathMax JavaScript library, which is a fantastic resource for making math work in a Web Browser.

In the standard user page on any Geometron system you will start by seeing the home scroll. If you are on a screen which is wider than it is high(like a laptop), there will be a bar on the right side of the screen with a list of scrolls and maps you can click on. On a device which is taller than it is wide, like a smart phone or tablet in their normal orientation, there will be a button in the lower left corner to "show menu" that you will need to click to see the lists of scrolls and maps. Scroll through the scrolls and click on them to read them.

To edit the scroll you are reading, click the edit icon, which is denoted by a pencil. This will drop you into the editor page, where you can edit the code for the scroll you were just viewing using the Markdown language as mentioned above. To switch which scroll you are editing, click on any of the scroll names that appear next to the edit area and you should instantly switch to one of those. To start a new scroll, enter the name of your new scroll in the "new scroll" input. To view the scroll you are editing at any given time, click the link with the scroll icon, which looks like an ancient scroll with two rollers at each end like a traditional Torah scroll and you'll see that scroll, then when you click the edit icon again you will get back to that scroll. If you click the Home button you get back to the home scroll regardless of what you were editing. The red "x" is the delete scroll page, which we will deal with shortly.

Copy often! We do not "save" in the normal sense in Geometron. Every keystroke edits the permanent file on the server. But all of that can be destroyed in an instant by any user at any time. When we want the documents we create to replicate we simply replicate them: copy all the text from the editor window and paste it into both other backup scrolls on the same system and to other scrolls on other systems, local copies on private machines, or global copies on public pastebins as will be discussed in detail below.

As stated in the Organic Media chapter, one of the Laws of Geometron is that everything dies: everything must be easy to delete or destroy. We delete scrolls by clicking on the red "x" to get to the delete scroll page. This page is pretty self-explanatory. Delete things!!! Everything is supposed to replicate, so get in the habit of destroying information as you go, and replicating the good stuff rather than holding onto a stale system of information which slowly rots over time as is the case on private hard drive in the property-based systems.

The first thing to understand when formatting scrolls is how you make a paragraph break. To do this you want to hit return not once but twice at the end of each paragraph. It is easy to accidentally only hit return once and that will run paragraphs together without any break when they are displayed.

The next most important thing to know is how to make a heading. This is done with the number symbol "#". The more number symbols, the *smaller* the heading. Thus one number symbol is the largest size, and when you want a sub-heading you use a double number sign and so on. Headers will be centered when displayed.

To emphasize a bit of text, put it between asterisks. To make a bullet list of items, put a dash before each item, with a space after the dash. To make a list with numbers, do the same thing as with the dashes but with numbers followed by a period. Again, put a space after the period to tell the system this is a list. Be sure to put something after each number or bullet in the list, or the last one will end up looking weird.

Dealing with images is a little bit different than what you might be used to on other non web-based systems. As with the rest of the Web, images are put into documents as links to a location where that images is stored. That location can be locally on the Geometron server you are working on but ideally it will be on a globally visible web address. The simplest way to do this is to find an image of the thing you want an image of that is already on the free open Web like on Wikipedia using a search engine, and then right click on that image and copy the image address to the clipboard of whatever machine you are using, then paste it as the address of the image. Images are put into Scrolls using the somewhat confusing Markdown syntax which is an exclamation mark, two square brackets, and then parentheses containing the link. Yes, it's weird, but it's not hard once you get the hang of it, you can just copy paste from existing images and drop you new image address in.

We can also upload images to each Raspberry Pi server for documents which we are just working with locally, and this is useful when you want a locally specific image which is of no use outside that server like a photograph of a physically local place near the server. This is done using the Local Image Feed, which is documented in the Feeds chapter. The other main work flow we will use in Geometron is taking original photographs with mobile devices, uploading those photographs to publicly visible locations, then copying the address of those images and pasting it into scrolls. One way to do this is with the website Imgur.com, which allows people to upload images both with an app on a phone and over the Web on a browser. You do not need an account to do this in a browser, but will need one to use the mobile apps, but it is free to use. Set the privacy of the image to "hidden" and then always copy and paste that url somewhere. Also note that there are several ways to share a url of an image, and you need the one that links to the raw image, which should end with some kind of image file format like .jpg or .png. To get this the most efficient thing is to right click and copy just as with public images we find in the Web at large(copy image address or copy image link depending on the system you are on).

The last but perhaps most important part of Markdown syntax we need for making and editing Geometron Scrolls is links. A link is created with square brackets containing the text of the link followed immediately by parenthesis containing the link address. For example we might make a link to the Geometron home page which is at index.html with [home](index.html). You can link to anyplace on the Web by navigating a browser to the page you want to link to and copying the address from the address bar and then pasting it into the parenthesis of your link. Be sure to always have no space between the brackets which contain the text and parenthesis which contain the address or the link will break. Also, links can be images. You can put an image inside the square brackets and that image will appear as a link to whatever is in the parentheses like any other link.

Geometron scrolls allow for local links to scrolls and maps in the Geometron system in addition to the global web links described above. All scrolls created on any given Geomtron server are stored in the "scrolls' directory. If you put a link to "scrolls/scrollname" the Scroll software will detect that as a scroll link and clicking on it will load that scroll into the scroll reader rather than linking to the raw scroll file. Maps, which we will discuss in a later chapter, are linked to in the same way with links that start with "maps/". If you want to load a scroll or map stored in a remote location into the scroll reader you can use the scroll() or map() commands in the link, with the destination inside the parentheses, which are in turn inside the larger parentheses of the link. For example, you might put "[link text](scroll(scroll location link))".

Another important aspect of linking scrolls is the user of the active user page called user.php. PHP is the universal web language we use for any kind of active program which engages the files on the Geometron server. PHP allows users to send information to the program which loads the web page they see by adding bits of information to the web address the browser points to. User.php takes two inputs, "map" and "scroll". To make a link to a scroll called "trees", we would construct a link as follows: "user.php?scroll=scrolls/trees".

Similar to user.php is copy.php, a universal copy program which can copy a file from anywhere on the Web(really anywhere) to anywhere on the Geometron server you are working with. Copy.php takes two inputs, "from" and "to". When a PHP script is fed inputs, the second one uses an ampersand and the first uses a question mark. The order does not matter, but by convention we start with "from". So to copy the Scroll "tree" to "tree2", we would create a link to copy.php?from=scrolls/tree&to=scrolls/t

It is worth mentioning that you can always use HTML inside a Scroll and that will work, if you know HTML you can use this to add interesting things. Learning HTML is a useful thing to do, and dropping little bits of HTML into a scroll is a good way to do that when you're starting out. Also, you can use the code editor at editor.php to edit the main file index.html and change the default style of scrolls if you want to. As with HTML I won't go into detail on this, but I recommend that people learn CSS if they are interested in getting more into development and then it is pretty self-explanatory how to change the settings at the end of index.html.

Another important part of the work flow is the use of paste bins. You can copy a scroll into a paste bin at pastebin.com without getting an account or logging in. Set the paste exposure to "unlisted". You can also set the paste to expire after a certain time, which can be useful. When you paste a scroll into a pastebin, you will want to get the address of the raw file, by clicking on the link which says "raw". This is important! If you link to the first page you see after posting a pastebin the use of that pastebin in Geometron will cause problems. The address should actually have the word "raw" in it if it is correct.

The raw pastebin address of any scroll on the Web can be loaded on a local Raspberry Pi based Geometron server by putting the address into the text input that appears directly above the list of scroll names and the edit icon in the scroll list. Remote scrolls can be copied locally and edited for further sharing of new versions using the program pastescroll.html, which should be linked somewhere on the home scroll of your Server. This program lets you enter the pastebin address of a scroll, which can also be the address of a scroll stored on a globally hosted Geometron server or another Raspberry Pi elsewhere on the local wifi network, as well as the name of the new local scroll you want to copy that global scroll to. When both of these inputs are filled, an edit link will appear on the screen, and when you click on that the system will copy the remote scroll to the local file and you will be editing it. You can immediately edit it, then click on the scroll icon to see it, then back to the edit screen and so on, all as a local version of that scroll. Having edited that scroll to make it better, you can then copy and paste it from the scroll editor whatever Pi server you are on back to another publicly visible pastebin which you can then send the address of to another person anywhere in the world. They can then repeat this whole process on their local system, creating a copy of your new version which they edit and improve upon and then send the new link back to you and so on. This workflow enables global collaboration very quickly between people all over the world, all working on physically local networks that are not visible to the outside world, in particular search engines.

Note that public pastebins from Pastebin.com have censorship which can be somewhat random to predict, and they will prevent you from posting scrolls that have certain political topics. To avoid censorship filters, you can encode the document in its ASCII/UNICODE by using the program textconvert.html. This will let you convert any text file to a list of base 16 numbers separated by commas which can then be transformed back manually. You will only occasionally need to do this, and it is somewhat annoying, but it's useful to have as a tool. It also lets you deal with non-English characters which get turned into UNICODE, making it easy to copy and paste in English-only computer systems while preserving the information in non-English text documents. This should also be linked from one of the main scrolls in every system.

Pastebin can also be used to copy to global servers.

This is done with pastebin.html. This is the fastest way to clone a scroll to the home scroll of a globally hosted server. This will not work the first try, usually, you'll need to reload a bunch of times at random for the copy to actually work, for unknown reasons. In general, using Geometron on paid hosting platforms is spotty and you just have to try a few times to make it work. Eventually, when Geometron has its own servers hosting domains we can fix this, but for now just poke at it and keep trying and it will work.

It is difficult to overstate the versatility of a text document with formatting like this. The number of potential uses for a freely shared system of text documents without property vastly exceeds what I can think of in a book like this, but I'll say a few words here about what I have in mind for it.

The primary use of the scrolls which was the initial motivation for creating this system is for replicators, as with all elements of Geoemtron. This whole system is built on the idea that the most fundamental thing we can do in media is communicate how to copy a thing built from trash and other found materials in our environment. The basic workflow here is to photograph every part of the assembly process, link to all the links of materials we might need to buy(yes, we still have to buy things for now), describe steps, and link to scrolls and maps of related things.

When Geometron scales up, this will be a vast network of interlinked documents which connect people with things. Most people will not make most things, just as most people do not make most things in today's consumer society. But the things we use will always be *linked* to their means of replication. This means the physical thing contains the information in some form to *find* someone who can replicate that thing, and that that person will be able to *find* all the materials and learn all the skills. Replication means that this universe of linked documents has to be fractal in the sense that when someone chooses to dive deeply into a thing they learn not just a set of instructions but are linked to all the deeper knowledge required to attain expertise in the technical disciplines required for replication. This means that for example every electrical device should have links from document to document which lead an interested party into a whole curriculum of learning about electronics theory and practice to become competent as an electrical designer and builder.

Scrolls can also play a role in *finding* the free things we create and replicate. This is an important part of any economy: making a web resource that tells a user how to find a thing in a place. This is in some sense the majority of what the consumer media does with their constant barrage of advertising: tell a person how to find a thing in a place. We can do this much more effectively by pointing people to streams of things which are made on location so we are not just pointing to free things but to a free *stream* of things.

Also, a scroll is just a document. So it can be any document! You can write an article about...anything! Write news stories about the street corner where the Pi server is located. Write editorials about the current pastry selection at the local coffee shop where another server is. Write manifestos. Write the Constitution for the new system of local governance which applies only to people on your local wifi network. Write guides to the edible plants to be found within walking distance of your local Geometron server. Write your life story. Write an ad for whatever it is you do for money to survive our current money-based system. Create a long list of links to all your favorite obscure web pages. Create a recipe with details on how to find the ingredients in your local grocery store.

The use of Scrolls for sharing information locally can be incredibly powerful. We are building local media complete with news articles, history, politics and so on that are physically local to an area limited in size to what you can hear, see, or connect to over wifi. This can be used to organize workplaces against the central bosses who control the workplace from thousands of miles away, in everything from an Amazon warehouse to the drivers in a rider pickup area by an airport. We can use this platform to bring out a sense of place that has been lost in our current civilization and to share that place with people in a deep and complex way. Whole epic histories can be written of a street corner under a bridge, of how that bridge was built, of who has passed through there, of the bus routes that serve it, of the origins of trash that turns up there, of the destination of the storm sewer that drains water from it. Myths can be written, whole new religions built up, all around objects found in our immediate environment.

Here we recall the Laws of Geometron: everything is physical, everything is fractal, everything is recursive. We are building a network of documents which refer to physically local places where that physicality matters. We also make this fractal, given how much information can be on a server(hundreds of thousands of scrolls can easily be on a single Pi server), so that we can have media that fractally zooms in from a street corner to a trash can to just the plastic bottles that go through that can. And it is recursive, in that our documents are constantly referring back to one another. No document stands alone, they are all always pointing to other documents, just as is the case on the Open Web, but in a physically localized way and without any users, just free un-owned documents.

Physically local media can create value which people are willing to pay or barter for in the existing system in exactly the same way that consumer media creates value. A media platform serving a strip mall can provide interesting and useful content to the workers and customers and passerby, but because it can stimulate commerce, it will also create value for the business owners who will then have an incentive to contribute by barter to support the Operators of the network, who can in turn provide custom scrolls and links to them promoting whatever those business owners are doing.

This is also true for the informal economies of local places. The ability to have a mobile and compact media platform that can represent a place but not be obvious or permanent allows vast amounts of commerce to happen in a physical place without being visible to the rest of the Internet. In some sense this is a kind of "dark web" even though it is totally free and open and unencrypted. Having no property or usernames or passwords or cryptography means there are no names of people of any kind attached to the documents. This creates a total freedom to exchange goods and services in a physical locality with a physical trust model based on actually talking face to face with people and directly exchanging things and media.

To see how this can have a high impact, we examine Craigslist and why it's awful now. Craigslist is now all scams and spam for numerous reasons. Essentially, it got too big. It was unable to really become fractal, and there is no local control. Because it is centralized, even though they have separate boards for cities and for areas, those areas are still huge, and they are just theoretical: anyone in the world can see them and post, even though they're far away. Also, an area might have millions of people in it and cover tens of miles. Furthermore, central control means it is not tenable to have human operators involved with document creation, so spammers can spam relentlessly until everything is ruined. Even though posts are quasi anonymous, they still all have users, passwords, and control. If bad things are posted, the vast majority of users are not empowered to delete them, so if a whole board is ruined by a few spammers, no one can fix it and it stays ruined. In an open document model, a spam-filled board will get cleared out instantly and if it keeps getting spammed the whole board can be permanently destroyed, and all useful documents cloned elsewhere. Craigslist wars to "deal locally" for safety but again they mean this on the scale of a vast area with millions of people in it. Our model is more line-of-sight: you deal with people you can see, in a place where people you trust can see both of you, out on the Street Network.

Finally we must briefly discuss the workflow for technical documents which use math typeset in the LaTeX system and how to use that to create documents like this book. The actual details of this typesetting system are not covered here, as that is a very deep subject. There are numerous print books and online guides to learning it. The very short version is that you use dollars signs to enable math mode, and lots of back slashes and twiddle brackets and keywords to create arbitrary math formatting. And it all works in all web browsers! This is thanks to the magic of the MathJax JavaScript library mentioned above.

Math scrolls can be viewed along with math maps using mathuser.php, which is used the same way as user.php listed above. Editing is still the same basic method. Tex files in the home directory of your server can be edited using texeditor.html, on which this book was mostly written.

Chapter 6

Feeds

A Feed is an ordered list of pieces of information. This structure of information, in which a set of media elements like images or small bits of text are related to each other by order, which a person can scroll through, is the basis of most social media today. What we are doing is so radically different from anything you have ever encountered, however, that it is worth taking some time here to explore how exactly these feeds work in consumer media and how differently they work when we move to Geometron.

In a consumer-driven social media system, everyone involved is either a user, a worker for the company who owns the platform, or an advertiser. In order to interact with the system, a user is required to have a unique identifier based on their user name and password. Every single thing they do, from how fast they scroll to what they click on to every post they make is tracked and analyzed by the employees of the company. Advertisers then pay the company for the ability to spy on and manipulate those users into consuming more, and it is the job of the company employees to do that as efficiently as possible.

The flow of information in these consumer systems completely structured around these separate user accounts everyone on the consumer side is forced to use. Even in supposedly anonymous forums, posts are tied to IP addresses and can't be edited or deleted by other "users". That is to say, in the consumer driven systems, even without names the basic paradigm of the individual user is dominant.

Each user creates a stream of information which is "owned" by them, which is "their" feed. When each user is on the system, however, they see another feed, also called "their" feed, which is the sequence of information the company chooses to show them. This is a mixture of elements from other users' feeds and all the media being forced on them against their will by the advertisers who are paying for the network. In order for all this to work, the company has to give them just enough elements in their feed that they keep staring at the screen, while feeding them as many corporate manipulation messages as they can get away with to maximize shareholder value and profits. This has to violate the users' consent or it doesn't work–the whole system has to involve a central power who can control what everyone sees or people would simply turn off the feeds of the predatory manipulators and the system would collapse from lack of financial support. Also, the existence of "users" is needed in the current system again to maximize profits, because that is what allows the manipulators to target based on detailed profiles of who clicks on what and more importantly who will consume what.

Users are encouraged to have things "go viral" but always within a controlled environment where the company can shut it down and censor it and where each copy is linked to a user so that again the users' behavior can be carefully tracked in order better to manipulate it later.

Geometron has feeds as well. We have feeds of images, feeds of text, feeds of symbols, feeds of icons to be printed by robots, feeds of hyperlinks, and various feeds that are complex combinations of these. Our system of creating feeds is so simple that anyone with a very basic command of web development in HTML and JavaScript can create their own feed applications within an hour or two with copying and pasting of existing apps. In this chapter we will examine how our feeds work and how they differ from consumer feeds and then talk about the existing feeds we use in this system, largely to connect together other parts of the system. These feeds are potentially very powerful, however, but it remains to be seen exactly how they get used. Try stuff!

In Geometron, we do things so differently that it can take some thought to to see just *how* different this network is. We now recall some of the relevant Laws of Geometron and discuss how they impact what we do on this system.

No property. Again, while this sounds radical, it is in fact much more radical than it sounds. There are no users on this system. No employees, no company. No "core developer team". No "marketing customers". There is no user data because there are no users. There are no passwords, there are no logins. The servers themselves are non-property objects which exist to be shared freely from place to place. In our system, feeds are just documents. They are always in a pure text format which can be interpreted by the browser and turned into something readable.

Everything is replicates. A feed can always be copied and pasted in pure text(using the JSON format). It can be stored in a public pastebin just like scrolls, or stored in a text message or email or embedded in a scroll. A feed can be copied instantly from one server to another, spreading the whole feed as an object, without any regard for where each element of the feed came from. Everything we create is designed to replicate.

Everything dies. Every element of every feed can be deleted instantly by anyone interacting with that feed at any time. This is how we create a system where replication only happens with the consent of the people.

Everything evolves. Every feed can be edited by anyone at any time. There are no "permanent" files. To share, read, or copy is to be able to edit.

Everything is physical. The structure described here might sound unimaginably chaotic compared to existing systems where all information is controlled by a central authority and by private user accounts under some type of algorithm. But that is because we are used to global networks, where distance and geography don't matter. Our system is physical, centered on the community in direct proximity to the physical server. This means we are only trying to organize information in a way which is useful to the community physically surrounding this server. This makes the information sorting and organizing vastly simpler than on a global network. This limitation in the who and where parts of our system is important partly because it is what makes our model work in which everyone can edit anything at any time. This only works if people don't edit the same exact piece of information at the same exact time. Because we are all effectively in the same physical room, but are assumed to be wandering in and out, this is a dynamic which can be built up socially in an organic way.

Everything is fractal. Again, this is how we can make sense of the potentially overwhelming complexity of a feed without users. As soon as a feed becomes too complicated, we just fork it into multiple feeds based on sub-topics. Part of the chaos we are managing here is the potential collision of edits referenced above. Again, this can be resolved with a fork. The program fork.html will create any number of sub-pages below the tcp level, where as we add more fractal divisions we can avoid edit collisions.

No money. We do not have "customers" who can simply buy their way into forcing people to see their information. While network Operators might take donations for teaching people the system, because everything people don't like is deleted, there is no way to enforce the contract structure used in consumer society to pay for advertisements. Money plays a role in our system initially as we all need to survive somehow, but is not the backbone of the system like on a consumer system. Also, consumer systems require constant money flowing in in order to keep the server farms running and all their employees doing work they hate to manipulate people. Our servers are passed around through a community in ways of direct use to that community, so no global cash flow is required to keep them running in the long run.

While I am trying to keep this discussion non-technical, it is necessary to say something about the format used for Feeds in Geometron. Geometron feeds come in two varieties: a directory with files in it, and a JSON file. JSON is a format which stands for JavaScript Object Notation, and which is a very simple and universally recognized text-based format to organize information. Any of the hundreds of programming languages in common use today will have already built in routines to handle JSON, and any programmer on any system should already be familiar with it. Like Markdown for the Scrolls, we select JSON because it is easy to copy and paste, and is as "lightweight" as possible, meaning we need very little added information in the form of weird looking symbols and so on to store things.

The most basic type of JSON structure we use is the array, which is just a collection of things separated by commas, inside square brackets. The other main thing that exists in JSON is the "object", which is a collection of pairs where one element of the pair is a piece of information and the other is a name for that information. This idea of creating abstract objects which map names of things to things and organize them in this way is part of the "object oriented" idea which is the basis of most modern computer systems. Never be afraid to edit and read the raw JSON! You don't have to interact with it but you should not be afraid of it. If you destroy a JSON file, the file you started with can just be replicated again and again to avoid the fault.

The first feed we will discuss here is Chaos Feed, which is perhaps the most basic, and is really just designed to show the concept. Chaos Feed can be found along with the other feeds from the Feed Scroll which should be linked from your Home Scroll on whatever Geometron system you are using. To post, type and hit return. That's all. To delete, hit one of the red "X"'s. If you are using a keyboard, the up and down arrows scroll through the existing elements. To clear and start over, click the button in the upper right. To reload click the icon in the upper left. To see and copy the current feed, navigate to data/chaosfeed.txt on any Geometron system and you will see the JSON for that feed. To copy a feed from a remote pastebin, use copy.php as with the scrolls described in the previous chapter. To edit a feed or to edit the application chaosfeed.html, use the code editor editor.php. From editor.php you can edit any code on the system, including both the file chaosfeed.html and data/chaosfeed.txt. While most people will probably not want to edit this code, it is simple to edit for people who know basic web development, and so you can ask around and find people who like to do that to evolve the system. I do not know how people will use this. Try stuff!

Another feed almost identical to Chaos Feed but with different applications is urlfeed.html. This is the same interface: you put something in the input and hit return and it posts. But this is for links. Each entry turns into a live HTML hyperlink which actually links to whatever the link points to. To use this, find a link either locally on your system or globally and just copy and paste it in and hit return. Edit, delete and share just as in Chaos Feed. This feed unleashes the power of Hypertext, the central technology of the World Wide Web. Connect from your humble local feed to the whole world! Connect your system together. Connect everything.

Yet another feed of almost the same format is the Global Image Feed, or globalimagefeed.html on your server.

In this, you again just paste links in and they post, but in this case you are pasting image links which will then appear in the Feed. Again, these can then be deleted. Click on any image to see the url for that image appear in the text area. This Feed links up with the alignment system for tracing Icons which will be dealt with later in this book. But it can be used for anything, including just sharing random images from around the Web or use as inputs for any other Geomtron application. We will deal later with a number of such applications which can call on this feed and use it to load useful images to do things with. Once again, delete anything at any time you want by clicking a red "X".

I said above that there are two types of Feed, and we now turn our attention to the second one: files in a directory. We use this for images uploaded to a server as well as symbols created on a server. Images can be uploaded to a Raspberry Pi Geometron Server using localimagefeed.html. This is also used as part of the workflow in several other part of the Geometron system. Note that there is a maximum size on images, so it can be useful to screen shot images on a mobile device, then crop the screen shot to get something closer to 1 megabyte or smaller before uploading. The maximum is about 2 megabytes. This is to keep us from having exploding sizes of data based on very high density images. This is a key feature of our physically local social media system! We can use this to photograph objects and places in direct proximity to the server and upload them and use them in order to build media which directly represents our physical environment. To upload, click "choose file" first, and select a file to upload from your device, then when that file is selected use "upload image" to upload the image. To delete, use the red "X". All you are seeing here is a list of the files in the directory uploadimages/, which you can manually examine and interact with as well. This can be an easy way to transmit images and memes in a peer to peer way, as a meme or image can be uploaded from a mobile device to a local server, which another user can then download to their mobile device, allowing for rapid peer to peer transmission locally over the wifi network. As with the other Feeds here, this is deceptively simple, and can by itself form the basis of a whole new type of local social media networking.

The real heart of the Geometron system is the geometric programming language presented later in this book. Among other things this creates image files which are symbols in either the vector graphics format .svg or the bitmap format .png. These are all stored in the directory symbolfeed/. They are viewed and deleted using the feed program symbolfeed.html. This whole system will be explained in detail later in this book, but for now it is just important to know it is there: this is geometric social media, in which you create geometry and share it freely. Again this is deceptively simple, and can form the basis of a very wide range of technological activities which will be described later(along with the Icon Feed, stored at iconfeed.html).

The Feeds described here are programs which run on a Geoemtron system. We must also consider Feeds in a more abstract sense which we will use to build up our whole system. Ultimately the whole system will be based on the Trash Feed, the vast global feed from mine to landfill which we will redirect toward our new civilization. This feed is made up of many small local feeds, which have a fractal structure. For example, a trash bin near a kiosk in a park selling bottled water will be a constant source of plastic bottles. This type of feed is part of our system in a general sense even though it is not software. But it is not totally separated from our software either. Our system includes robotics which can print icons into plastic bottle caps with a heated tool, and those icons are shared on the Icon Feed, so there is a crossover between the physical feeds of trash-sourced objects and this less tangible software system.

Chapter 7

Maps

Maps are a from of document in which a set of images, words, and links are arranged geometrically on the screen. Just as the Geometron Scroll can be thought of as a replacement for Microsoft Word, the Gecmetron Map can be thought of as a replacement for Microsoft PowerPoint and Keynote. But it is also a way to make and share memes, to annotate geographic maps, annotate photographs of objects and really do any kind of communication where the relative geometry of objects matters.

It is worth once again examining the consumer civilization's version of this in more detail to understand what we are trying to do differently here. PowerPoint is the language used in the replication of things in today's world. When a new company is born, the founders use a PowerPoint slide deck to sell that company to investors. Every government applied science project starts with PowerPoint slides(often just a single one in the infamous "quad chart" format). In many ways our whole civilization runs on PowerPoint today. It is hard to imagine any project being funded in the world today, be it government, corporate, or non-profit, without a series of these simple graphic constructions of text and images and vector graphics arranged in a geometric order of some kind. And of course this format is the basis of the meme, this bizarre new type of thing that spreads freely across the web, generally with copy and pasted bitmaps.

We also need this format for replication of technology in our trash-based civilization as well as for all kinds of other communication. However, we need it to fit in with the values and laws of Geometron, and that requires that we rewrite the whole thing from scratch.

As with every Geometron document format, this format has to be something human readable using plain text so that each individual Map can be pasted into a text message, email, or pastebin for freely sharing across the Web without any intermediary. We do this using the same language as the Feeds discussed in the previous section: JSON(JavaScript Object Notation). A Map is an array of objects, each of which has a collection of properties. The array is denoted by a pair of square brackets, and each element is separated by a comma. Each element is inside a pair of twiddle brackets, and consists of pairs of names of properties and values of those properties. Each element has a position, a size, an angle, a text value(optional), a link(optional). an image(optional), and information on whether it is a global link or a local link inside the Geometron system.

You don't need to understand what all that means to use the system! The technical description is just there for reference. Maps are edited and drawn using a JavaScript library called mapfactory.js, which is replicated with each instance of Geometron. If you are a web developer feel free to go read the source code now to get an idea of how it works(there is not much to it).

By default, maps are displayed in a square area on your screen, and when you load the Geometron home screen that square will be either the left or top part of your screen depending on if your screen is wider than tall(landscape) or taller than wide(portrait). To load a map, just look at the list of maps in the menu either to the right of the screen(for landscape) or in the popup menu you click to open with the button and look at the right side list and click on any map. That will load it. Now you can try clicking around on all the maps on your system, as well as navigating from map to map using internal links inside the maps(some have this some don't).

Maps are used to amuse, to tell stories, to make points, to denote where things are, to point out where a part of an object is located, and to connect web pages to one another. Maps are much more powerful than the Power-

Point slides they replace for several reasons. The ability to have both global and local links to other documents makes them fully integrated into a global, ever-evolving network in a way that makes them much more rich and complex. They are, like all Geometron documents, not owned by anyone. Each individual map is a free document, which can be replicated, edited, deleted, shared an infinite number of times instantly by anyone on the system. This creates a richness of information which is impossible with a dead file format like PowerPoint.

Maps are a great way to build social media around a physical place. When community forms around a local server in a local place, the local media should have photographs of the objects in the environment. Unlike a consumer network made up of "users", we have people in a community who share documents openly and freely. Photographs of people in our community can go here, and we can build up a media pool that includes us all, but as a shared community of documents, not as a database of "users". It is also an important way of denoting the exact physical location of things if we are to build up complex systems of industrial production from trash in our immediate environment.

Maps are also another way to rapidly create web pages in the Geomtron system, as the main home page can be set up to point to them directly by changing one line of code in index.html, by replacing loadscroll() with loadmap() and the name of a map. All maps are stored in the directory maps/ on each Geometron Server, and you can see all the maps by looking there in a browser, then click on them to see and copy the raw text of the map. Just as we have a home scroll stored in scrolls/home, there is a home map stored in maps/home.

Maps are edited with the map editor, which is at mapeditor.html on your local Geometron Server. The Map Editor looks a little bit different on a portrait versus landscape screen. In either case, the screen is divided into different areas which have links and buttons to do different things. The main map window should look the same as when you are in the passive map reading mode: a square either in the left or top of the screen. The element edit window contains icons indicating various actions, including select next/previous element, move element up/down in list, delete element, create new element, save map, delete image, delete link, and selectors to select which type of object you can select from to replace in the element you are editing using the textfeed described below. It is a good exercise to just try playing with these, deleting all elements, then making a fresh one and playing with that. Whatever element is selected is moved by dragging around on the map display screen, which can be done either with drag-click with a mouse or touch-drag on a touchscreen. Elements are resized or rotated using the zoom/rotate box, which contains slider bars for both scale and rotate as well as buttons for both zoom and rotate which only appear in landscape mode.

There is another window which lists all the maps, and you can click on those to select which one is being edited. That window also contains links to the home screen, a link to the specific map you are editing, and a link to the map delete program. The Map destroyer is exactly the same as the Scroll destroyer. It is a list of all maps on the Server, with a button to delete each map instantly. There is no undo. To click is to destroy. If a map is worth saving it is worth copying and sharing and if it is copied, deleting it costs nothing, so we make it very easy to delete. The destroyer page has a link back to the map editor.

The window with these links also has a text input where you can input the name of a new map. Try entering the name of a new map, and you will see a blank screen. To add an element to that map, click the icon with the plus sign. To delete it click the red X. Add another one, and move it to the top then bottom, move them around. To add an image to a map element, click on one of the images in the window with the images. Then you can click the icon with the red X through the image symbol(mountains and a sun icon) to remove the image and go back to just text.

That scroll of images can also display a scroll of links or text. All of these are taken from a combination of feeds, but primarily the Text Feed, at textfeed.html. This is linked via an icon with squares separated by a triangle. The local image feed is also displayed and you can add to that using links on here as well with "choose file" and "upload image". The blue link icon will make that scroll a list of links, and the ABC icon will make it a list of text elements. You can also edit all these manually using the table of inputs.

Unlike the Scrolls, maps are not instantly updated as you edit. They have to be saved with the save button. Whenever you save a map, the text based representation of the map in the JSON format is placed in the text area below the control buttons. If someone sends you a link to a raw map, you can go copy the contents of that map to the clipboard of whatever device you are using and then paste it into that text area and click the "import" button to import it into the current map. You can then save it, and the current map will be replaced by the imported map, destroying the existing map. You can also hit the "reset" button to clear out the map and start over with a fresh one. Try making a new map, then selecting all the text in the text area, and pasting it to a public pastebin, then sharing that link with another user. They can then paste it into the text area of their server to make a new map which is a copy of your map, edit it, and send it along to the next person and so on.

The button at the bottom of the table of inputs to edit the current map element after maplinkode sets whether that mode is true or false. If it is false, and there is a link from the map element it will be a regular hyperlink like any other link on the Web. But if it is set to true, it can point to either any map or any scroll on the web. If you paste a map in a pastebin and then get the link to the raw version of that pastebin, then put that into the "link" field in the table, you can click on that link and it will load that remote map from anywhere on the Web. This can be incredibly powerful, and can create entire networks of complex interconnected maps, all via anonymous pastebins, all referencing other images around the Web without storing any information on the local server or linking to any specific server or user(there are no users).

The button marked "height mode" changes the relative height of the element rather than the height and width together, which can be useful for sizing text elements exactly how we want. This does not do anything when the element has an image, however as the element will then automatically size around the aspect ratio of the image.

While you can enter the address of an image, the value of a text area, or a link value into the text fields, this is unwieldy and particularly annoying on mobile. Therefore the main way to insert one of these types of information into a map element is via clicking on it in the scroll of images, text or link. This is switched between these three by clicking the appropriate icon in the control button table(there is a link, text, and image icon). The image scroll is listing images from the local and global image feeds discussed in the last chapter, but it also has a special feed just for feeding information into the Maps. This is the Text Feed, located at textfeed.html. If you go to this page, you can input text, links, and images, and then go back to the map editor and use what you entered there. This is important for working on mobile devices where the manual input in the table is very awkward, but it is also important for another reason: dealing with symbols, icons, and other creations of the Geometron geometric programming language.

Creating symbols with Geoemtron will take up much of the rest of this book, but while it works with vector graphics, we need a fast and simple way to embed them in maps. To do this we use the so-called "base 64 encoding", which allows bitmap images to be inserted into files without reference to an external image file. Symbols made with Geometron can be converted into base 64 encoding using the link from the textfeed page which says "png code", and that will let you choose a symbol to convert and save into the textfeed, along with the ability to select how large that image should be in pixels. As with all Feeds in Geometron, you can delete any element of any of the sub-feeds at any time by clicking the big red X. Textfeed also links to some other applications which use it, such as the Duality and Poetry Engine apps, which you can explore as well.

Just as with Scrolls, we need to have the ability to use mathematical typesetting with maps. While most people may not need to typeset math, it is useful to know how to point mathematically minded people to these tools so that they can use them if they want. The starting point for this is to find a map or scroll that references math, and they should link to the relevant pages. The relevant pages are mathmapeditor.html, mathmapeditor.php, and mathuser.php. Combining math typesetting with geometric symbols quickly and freely and sharing them can build a whole new method for rapid communication of mathematical ideas. Setting up this system in physical proximity to a location with a community of mathematicians can be very powerful and is highly recommended!

Now that you know how to make, edit, delete, and share Geometron Maps I want to say a few more words about exactly how they can be used in helping with the workflow of the overall system. One of the most fundamental tasks we need to do to build a physical network on the Street is to help people to find things in the physical world quickly. Global map servers exist with maps of the whole world, which everyone can get to on any Internet enabled device. We can use these maps to find where we are, then screen shot the map, save it, upload it to a server, and build Geometron Maps which then annotate that map with symbols, links, and words to show exactly where a Server is, or where physical resources are for our trash-based industrial production. This can also be very useful for ad hoc mapping in an area with specific local maps, such as near the exit of a subway. Subway station exits often have a very specific localized map centered on the station with clear markings of relevant landmarks. If you photograph that, upload that photograph to a public image server, link to it, and annotate it, you can build a whole system of networked maps from that. Don't forget links! Maps can have annotations which are maplinks, linking to other maps or to scrolls with any kind of document you might want to associate with a place. A physical place can have a rich fractal structure of documents built around it, forming a kind of physical hypertext: a document in which we and everyone around us are physically immersed.

When we want to present a graphical story like the "pitch decks" which use PowerPoint in the consumer media, we can treat each Map as a slide, and then link them together with links on the edge of the screen going from previous to next and so on. This can copy the functionality of slide decks exactly, but with much more richness of content since they can also link to scrolls, do not have to be linear, and are not restricted to loading local maps. Rather than a "deck" being a dead document sitting on a private hard drive, linked Map sets of Geometron join a global swarm of potentially trillions of linked documents all replicating, evolving, and being destroyed in an ever shifting living informational universe.

When we build technology, and attempt to document that technology to aid in replication, we need to be able to label parts of the technology, and then link to images of sub-systems which then also have annotations. Maps allow us to do this in an infinitely fractal way, zooming in on things with more and more detail, and then forking off to related things which are linked in much the way that Wikipedia forms a vast fractal network of information.

The combination of geometric programming discussed in the next sections with Map creation allows for exploration of symmetries of the world around us which is unlike any other tool in its direct connection to the symmetries and scales of the Universe. This will be explored much more as we go along, but I encourage the reader to keep returning to the Map editor and making maps with all the other parts as we go along and learn the Geometron System.

Beyond these examples, there really are no limits to how much can be done with this format. It is lightweight, simple, easy to copy, and fundamentally more powerful than the predatory consumer systems like PowerPoint. The framework is so simple that interested programmers can easily rewrite the whole thing from scratch in many systems, making all sorts of applications with it. As with the JSON format itself, or HTML, the very simplicity of the specification is what gives it its power. A blank PowerPoint document with literally no content can still be well over a megabyte, and you can't read it or edit it without paying Microsoft rent for the "right" to run code already on your hard drive which they agree not to break if you keep paying them protection money. A Geometron Map can be as small as a few bytes, is human readable, can be shared by text message or email with direct copy and paste, can replicate itself freely across the network, and can be opened by an application that is itself only a few kilobytes in size. Build on this system! Make it yours! Share it and help it grow!

Chapter 8

Symbols

Symbols are geometry with meaning. The symbol is perhaps the most general idea that exists in human thought, since everything we do is mediated through the use of some type of symbol. When we speak of symbols in Geometron we mean *any* geometric construction which has meaning to people. This includes not written language like text but constructions like the layout of a microchip or the design of a building. It also includes the way we control machines, how we program them build automation. Geometron represents a new framework for working with all these kinds of symbols.

We live in a civilization today totally dominated by numbers and by the people who work with numbers. The machines we currently use to communicate are built by people who believe that the most fundamental task such machines can do is to work with numbers. They call all these machines "computers" and have a whole theoretical framework for understanding how to build them using the ideas of arithmetic. This works. But it is extremely inefficient and distracts from the real purpose of such machines. Do these machines do arithmetic? Of course. They also keep very accurate time, does that make them clocks? They produce heat, does that make them heaters? No. Just as we don't call a light bulb a "screw" just because it screws into a socket, it does not make sense to let the idea of the arithmetic engine dominate in a technology the sole purpose of which is to communicate with other people using symbols. At some intuitive level, most people understand this, it is why the smart phone is primarily called a "phone" rather than a "computer". But the underlying mathematical constructs which built the digital computer remain, along with a whole lot of mathematical flotsam and jetsam which have held back progress and made simple and free media out of reach.

In Geometron we are switching from a world view based on numbers to one based on geometry. This represents a shift in value system. In "computer science", the manipulation of numbers and logic are considered the most fundamental operations. In Geometron, we consider geometric constructions to be the most fundamental. This is a shift in perspective, which we can apply to the whole of the existing machines. How are these machines built? The microchips which make them work are nothing but huge geometric constructions, made up of little overlapping rectangles and polygons. These chips are then laid out on circuit boards which are again geometric constructions. The chips are placed automatically on the boards using machines programmed to carry out a sequence of geometric motions. They are packaged in cases made in molds again machined with this kind of geometric programming. And finally when assembled, their main task is displaying symbols on the screen which is again just geometric construction.

It is easy to forget given the onslaught of propaganda from Silicon Valley just how accidental the rise of their machines as the dominant technology was. We also are encouraged to forget that these machines were built *primarily* for war initially, then large authoritarian organizations to track and control people, and only later, almost as an afterthought, as the communication devices we rely on for all aspects of modern life. One of the theses of Geometron is that a shift in thinking from one based on numbers to one based on geometry is a shift away from the ideology of dominating large amounts of land and people toward one of cooperation based on sharing of technology and ideas.

When we build everything from trash found directly in our environment, empire-building doesn't really accomplish anything. The people 1000 miles away from you have the same piles of broken phones you do, so you gain nothing by dominating them and vice versa. But if you can share with them how to make those phones part of your free network, the value of *your* network infrastructure goes up exponentially, just as we find in all networks. as they scale up.

In computer science, they work with an idea called a Turing Machine, named after computer pioneer Alan Turing, which is a generalized machine for doing arithmetic. An infinite tape of ones and zeros is fed into this imaginary machine, and the contents of that tape give instructions to the machine, which then carries out actions on the ones and zeros on the tape. Any computer, regardless of the details of how it is built, can be shown to be equivalent to this toy model. In Geometron, we are creating a similar object: an abstraction which can construct any symbol, which takes symbols as an input.

The basis of geometric programming in Geometron is the Geometron Virtual Machine, or GVM. Just like the Turing Machine, this is an abstract construct which carries out geometric constructions based on a set of instructions. We assume that there is a main program, which we call a "glyph", which consists of a sequence of symbols, each of which represents a geometric action. Just as the Turing machine reduces every math problem to binary arithmetic, our machine reduces all geometry to discrete geometry. The GVM has an internal geometric state which represents its progress in doing geometric actions. If we are programming a physical machine like a pen on a plotter, the position of the pen is stored this way. This logic is very similar to that of languages like Logo, a teaching language developed in the 1970s, which uses what is called "turtle logic", where patterns are drawn by a virtual turtle which moves around on a screen, and which can also be performed by a physical robot with a pen. However what we are doing differs radically from Logo as we will see below.

We also have states of this virtual machine which describe what motions it can carry out. The most basic of these is the step size. By creating geometric programs using an abstract step size without actual numbers, we can create programs to draw symbols independent of what machine we use and what scale we are at, copying verbatim a program from a giant wall climbing robot which spray paints symbols on a building to a nanolithography system which prints the exact same symbol in an area smaller than a human hair, without ever dealing with the mechanism of either machine. This unit state is also used to define how we do constructions like "draw a circle".

The most basic geometric program we can carry out is the construction of the Vesica Piscis. This figure, from the Latin "fish bladder", is just two circles each of which has its center along the edge of the other. In the dialect of Geometron presented here, the symbols are all printed inside squares of identical size, patterned from left to right. The symbol describing the action of drawing a circle is a square with a circle in it, and that draws a circle with radius equal to the current value of the step size. If we draw a circle of radius equal to step size, then move sideways one unit of that same step size and repeat the action, we get the Vesica Piscis. The symbol to move to the side is just an arrow pointing to he side. These symbols are themselves constructed using more of the actions of Geometron: drawing line segments, rotating by discrete angles, scaling the unit down and back up and so on.

Another central organizing principle of Geometron is that there are special symmetries and scales that are intrinsic to the Universe which we use to simplify how we approach geometry. In a numbers-driven system, coordinates and angles are all equal. 37.34 degrees is no different than 36 degrees for example, they are just different numbers with no special properties. But this ignores some very deep patterns in how the world around us functions. Both the natural world and the constructed world of human technologies rely constantly on special rotational symmetries, starting with small numbers of rotations, in particular twofold, threefold, fourfold, fivefold, sixfold, eightfold, tenfold, and twelve fold. From there if we add halving angles and dividing them by 3 we can get all the way to the 360 fold symmetry which defines the degree of angular measure in most common use. Using only discrete geometric manipulations, we can go from fivefold symmetry which is based on 72 degrees, divide

116   CHAPTER 8. SYMBOLS

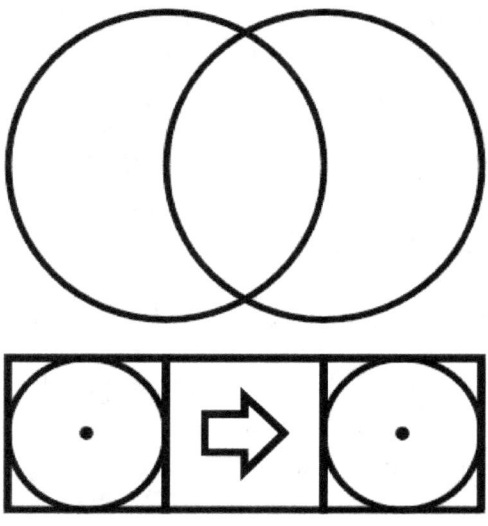

Figure 8.1: Vesica piscis, or "fish bladder" spelled out with Geometron symbol glyphs.

by 2 to get 36 degrees, then divide by 3 three times to get 1 degree. So discrete geometric actions can be used to do a very wide range of geometric constructions without ever relying on reference to numerical representations of angles. The dialect of Geometron presented is based primarily on 4,5, and 6 fold symmetry, combined with halving, doubling, dividing by three, and multiplying by three to construct all angular rotation actions.

By default, we increase or decrease our unit of step movement and construction by factors of two. Just like in digital computers, this binary representation allows us in principle to represent any number using only geometric actions of doubling, halving and moving by discrete amounts. Again this allows us to program a machine to go to any coordinate without actually using numbers, only using symbols which represent geometric actions. Those could all be represented by numbers of course, but we choose to *express* them using pure geometric symbols. And just as binary arithmetic can express any number, this binary geometry can express every position in space.

Another intrinsic geometric property we find in the world around us is the scales which naturally go along with these symmetries. For example, when we deal with fivefold symmetry, everything is based on the Golden Ratio. The ratio of the side of a regular pentagon to the distance along the cords which make up a pentagram drawn inside it is this ratio. If we then build a fractal of pentagrams inside pentagrams, the way we scale down to smaller and smaller pentagrams and pentagons is again and again the Golden Ratio. This number is about 1.6 and is a universal constant built into the structure of the Universe, found in all kinds of natural systems, as well as used throughout human art, architecture, and technology. While a numbers-based system can of course compute geometry using this number expressed as a repeating decimal, in Geometron we simply use a symbol

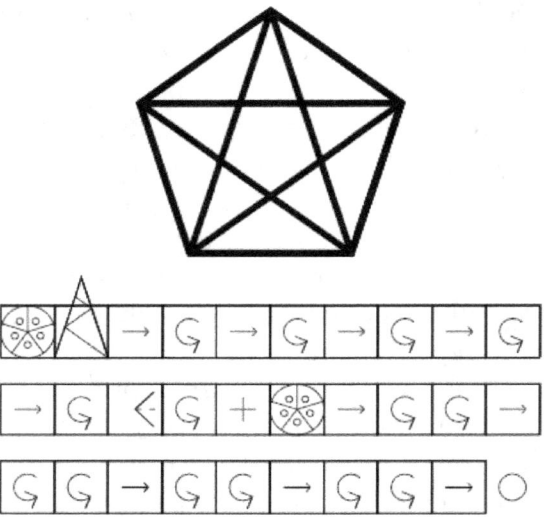

Figure 8.2: The pentagram in a pentagon, showing the relationship between the Golden Ratio and fivefold symmetry.

to represent this scale, without any symbolic reference to its numerical value.

The square root of three plays a similar role to the Golden Ratio, but for sixfold symmetry. If one connects alternating corners of a regular hexagon, those cords are the square root of three times the length of a side. Again, a fractal construction of hexagons and six pointed stars

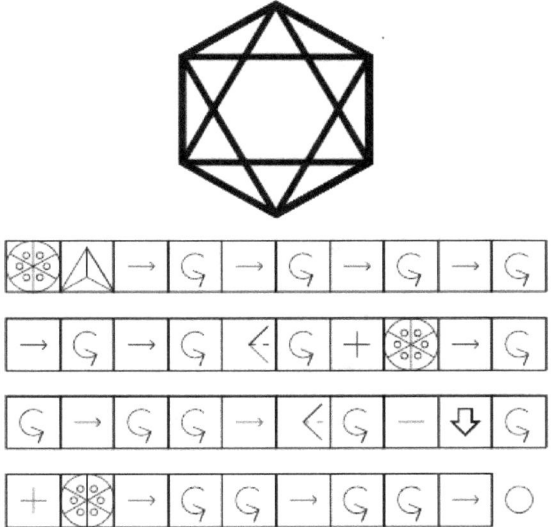

Figure 8.3: The six pointed star in a hexagon, showing the relationship between the square root of three and sixfold symmetry.

shows a square root of three scaling over and over. Similarly, the square root of two is intrinsic to four fold symmetry, as if we draw a diagonal line across a square that is the square root of two times the side length.

Altogether, the scales used in this dialect are, in order, the square root of two, the Golden Ratio, the square root of three, 2, 3, and 5. When a scale is set, that scale is the

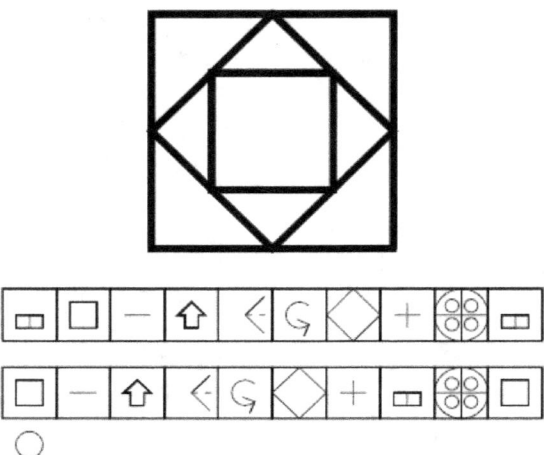

Figure 8.4: Embedded squares showing the relationship between four and eight fold symmetry and the square root of two.

factor by which the unit is either multiplied or divided when we apply a scale-up or scale-down operation.

The figures show the symbols for all these scale values. We are now ready to understand all 8 of the basic discrete movements: move forward, move back, move left, move right, rotate left, rotate right, scale down and scale up.

The current state of the GVM is expressed with the

Figure 8.5: Scale actions in order: square root of two, Golden Ratio, square root of three, two, three, five.

Global Cursor, a shape which shows the position, scale, the step angle size, the current direction which is "forward" and the current directions which are "left" and "right".

The basic constructions of Geometron are to draw a point at the current location, draw a circle of radius equal to the current unit, draw a line segment along the forward direction, and draw an arc from one of the cursor wings to the other. More advanced actions include writing letters, creating paths(both filled and unfilled, closed and open), and drawing Bezier curves, all of which will be covered in the next, more detailed section on the web-based graphics system which is built into Geometron.

Color and line width of lines are set with the layer system. At any given point along the construction of a Geometron glyph, one of the states is the current layer, of which there are 8. Each layer has a stroke color, a fill color, and a line width. These are set in an object which

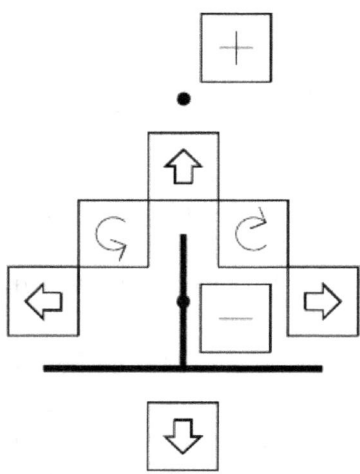

Figure 8.6: GVM cursor with movements. The arrows represent movement of the GVM position along the indicated directions relative to the cursor. Angle rotations are as shown. The plus and minus symbols are also shown and how the rescale to where the points are on the cursor.

Figure 8.7: Symbols for the layers have little line segments to denote the layer number, and show the line width, fill colors and stroke colors in the border and fill of squares.

can be edited and customized, which the GVM calls on when it draws symbols.

A very important point to make about how all this fits together is that each of the symbols shown here which represent these geometric actions are themselves constructed using this language. When a symbol is displayed in the spelling out of a Geometron glyph, each of the actions which compose the glyph has a symbol which is itself a Geometron glyph, and the whole sequence of symbols which spells out the glyph is itself one giant glyph which tells the GVM how to spell out this human readable symbolic description.

Geometron glyphs consist of a sequence of geometric actions. Each action has a symbol, which is itself made up of actions, each of which has a symbol and so on(recall that one of the laws of Geometron is that everything is recursive). Each geometric action is represented by an address in the Geometron Hypercube. The Hy-

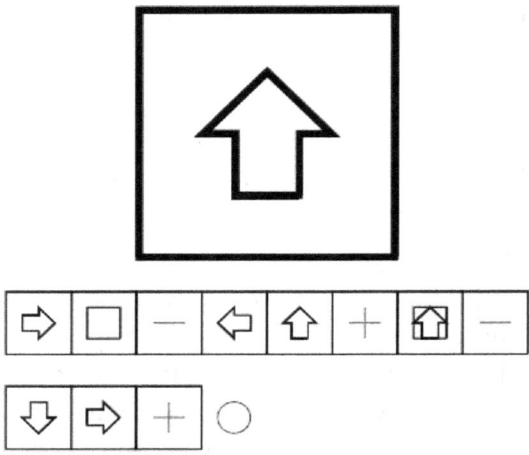

Figure 8.8: Breaking a symbol glyph down into its constituent action glyphs. This shows the spelling for the geometric action which moves forward one unit. It includes a sub-action which draws the arrow.

percube consists of two cubes, each divided into 8x8x8 = 512 cells, for a total of 1024 cells. Each cell contains a glyph, which is itself a sequence of addresses in the Hypercube. The Hypercube is therefore a kind of recursive data structure, with many components which all point back to itself. It might be added that human languages are all forms of recursive data structure, as they are described using the language itself(e.g.dictionaries). Hypercube addresses consist of four digits each of which is a number between 0 and 7, and the first digit of which is 0 or 1, just denoting what cube it in.

Why, you might ask, do we add this complexity? It is deceptively powerful to create a structure like this. Note that this structure is completely geometric. While we emphrepresent each cell with numerical addresses, the actual underlying structure is geometric and symbolic. Elements represent geometric symbols, human language describing geometry, computer language describing geometry, and locations in a data space. This is a nonnumerical construct, and represents a fundamental shift from the Turing model of computers to a model for a generalized symbol constructor. If our goal in building technology is to draw symbols, our fundamental models should reflect this, and not mask it in "computation".

Of the two cubes in the Hypercube, one is the "action cube" and the other is the "symbol cube", which has a leading 1 in the 4 digit address instead of 0. Each action in the action cube therefore has a corresponding symbol.

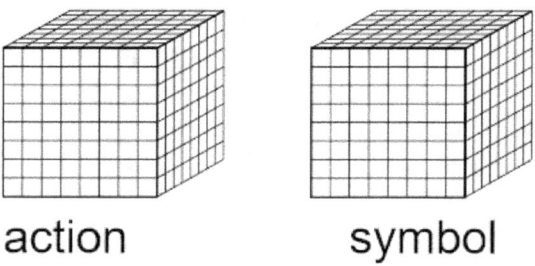

Figure 8.9: The two Cubes of the Geometron Hypercube: the Action Cube and the Symbol Cube.

Thus when a GVM is spelling out a sequence of symbol glyphs, it is just carrying out the actions represented by addresses of the form 01xyz, which are in turn simply doing the list of actions stored in that address in order. Any time you see Geometron symbol glyph spelling, you are looking at a sequence of symbol cube addresses.

But what of the Action Cube? This has a lot more structure than just all being actions. As said above, we are looking to build a geometric language which can apply to the widest possible range of generalized symbols. This means we want to be able to not only make 2d symbols, but to make complex fractal structures of symbols made from symbols for specialized graphics, printing in any human language, 3d constructions, editing the hypercube itself and perhaps most importantly machine control for generalized automation.

The Action Cube's bottom addresses from 0 to 037 represent actions directly on the Hypercube and the GVM and environment. These are used for tasks like moving the cursor around, choosing which element of the Hypercube is being edited, deleting addresses from a glyph, and changing the view geometry of the symbol display(zoom and pan).

Addresses from 040 through 0176 represent the printable characters on the standard keyboard using the ASCII code. These addresses all map to actions which are carried out when that key is struck on a keyboard. These are physical inputs. They are geometric in the sense that keyboards are geometric, and that hitting keys is a geometric action. The corresponding symbols stored in the Symbol Cube at 01040 through 01176 represent a font. These can be any kind of character, and can be used for any keyboard mapping to any human language. This represents an alternative to Unicode, in which each glyph is directly created using the Geometron langauge, rather than called from the system's interpretation of Unicode.

Address 0177 represents "do nothing" and can be empty or used as a dummy variable.

The addresses between 0300 and 0377 are the two dimensional actions described above and in the next chapter, which are used for making web graphics, saved vector graphics files, and all the symbols used to represent Geometron glyphs. These are in practice all some type of computer function call, and since computer code is stored in ASCII and ASCII is part of the Hypercube, these little bits of code represent the sequence of Hypercube addresses which can be represented as a sequence of ASCII codes which can map to addresses in the Hypercube, so we retain our generalized structure. Also, if we take this as a total abstraction, we could describe the geometric action using text in a human language, such as "draw circle of unit radius", and that can be encoded in ASCII to make it part of the Hypercube format as well.

The addresses from 0200 through 0277 are the "Shape Table" and all store sequences either in 02xy as well or in the range of 03xy. This is merely a convention, but these are used to build up specialized language, such as for building circuit diagrams or cross stitch patterns. This topic has its own chapter, as there is a very rich range of languages which we can build this way.

Addresses between 0400 and 0477 represent machine actions. The most basic actions for machines we consider are for machines that have 3 perpendicular axes of movement, of which we have several examples later in this work. Using only the actions of discrete movement and binary manipulation on the scale of the step, this system can be used to encode any motion of a robotic probe, but that only takes up one row of 8 out of the total of 64 possible actions. This is also a topic so rich in structure that it has its own chapter. The ability to use a web interface to do totally generalized programming of the geometry of machines for automation can revolutionize the control humans have over machines and over automation.

Addresses between 0500 and 0577 are another shape table, specifically for referencing machine actions in 04xy. As with all Shape Tables, this can be self-referential, and we can use this space to build up fractal structures of motions within motions. This can be a huge enabling technology for programming of automation using simple machines. The actions in this range are used to create a form of generalized "icon", which also gets its own chapter. Icons are sequences of movements on a rectangular grid, which either involve drawing a pixel or not. This sequence is used to create physical media using a variety of technologies, which are documented in the aforementioned chapter.

Finally, the addresses at the top of the Action Cube from 0700 through 0777 represent three dimensional constructions. These are abstract, and can be applied to any method of three dimensional construction. These also get their own chapter, and can be used to create graphics in the virtual reality format of x3d as well as .stl files which can be printed on a 3d printer. They are displayed in the browser live using webGL libraries discussed later on. The Shape Table at 0600 through 0677 is for calling actions in the 07xy cube, for building up complex fractal structures in 3d construction. This can server as the basis of a whole alternative system for creating parts for 3d printing, as well as a basis for web-based 3d graphics for both virtual reality and augmented reality.

The structures described here, the GVM and the Geometron Hypercube, represent a new way of thinking about how humans control machines. In this way of thinking, the purpose of machines is to encode geometric information on the physical world in a way which communicates information to another human being. We start with this as a goal and build up a set of abstractions to do that as generally as possible. Just as the Turing Machine toy model has been implemented in vastly different physical systems, our abstract idea about symbol drawing machines can take totally different forms which carry out the same task as well. This is a key element of how we will build the hybrid trash-based hardware architecture described later in this work. We do not want to compute things. We want to display symbols and images on screens, and control the movements of the machines we build from trash to automate building more things from trash. That is all. If we can display a generic geometric symbol on a scavenged screen, we can create fully Organic Media as described in an earlier chapter. That is, media which self-replicates openly on a physical network, which needs no mined materials, no money, and no property. If this is easy and uses trashed cell phones, the constant stream of broken phones will allow us to create a model of ubiquitous networking of Organic Media by way of the Street Network. With the consumer based system creating billions and billions of phones all headed to the landfill, we can imagine a model where there are screens everywhere, but they're not mobile because they don't need to be. In a network with self-replicating documents, you can read what documents you want wherever you go. You can interact with a universe of documents which flows through the physical world, as you traversethat world, interacting with and improving documents as you go as you interact with people in the physical world. The hardware architecture to realize this will be discussed in its own chapter on Full Stack Geometron.

133

Chapter 9
2d Web Graphics

The previous chapter introduced the abstract idea of geometric programming with the Geometron Virtual Machine and the Geometron Hypercube. In this chapter we dive into the specifics of how this is implemented on the Geometron Server so that we can create, edit, and most importantly share symbols from a web browser. The heart of what is described in this chapter is the Symbol app which exists at symbol.html on every Geometron Server.

In the Symbol app, we can build Geometron glyphs either using a specially marked keyboard or soft keys in a control panel. To learn the system, it is best to first get a keyboard and mark it up with the symbols so that you can see what you are doing as you learn. The examples and figures in this section give keyboard values along with symbols in many cases so that you can follow along if the keyboard is not marked and can get used to the layout. The layout is easy to change, and presumably will evolve over time, but for starting it makes sense to copy whatever keyboard layout the person sharing the system with you uses, in this case me. See the figure with the keyboard layout for guidance, and go ahead and put markings on keys with these symbols. I also encourage the further decoration of Geometron Keyboards, with extensive paint and modification so that it retains the distinctive flair of Trash Magic(rainbow paint and googly eyes). Paint pens are ideal for this, but nail polish or other permanent paints which can be applied with fine lines can all work. A less permanent solution can be found with masking tape on the keys and symbols drawn on the tape. Stickers can also be printed out and applied to keys.

The control panel of soft keys should be pretty self-explanatory: the buttons do the same thing as the keys on the keyboard and are there to make the system work with a touch screen when no physical keyboard is connected, as with smart phones and tablets.

With the ability to edit the main glyph, the best way to learn the language is to just try stuff. This chapter is largely pictorial. Go through the figures and try to copy them on the system you are using, play with all the different symmetries, scales, layers, and functions. To save a symbol, hit the save icon which is in the far left of the

136   CHAPTER 9. 2D WEB GRAPHICS

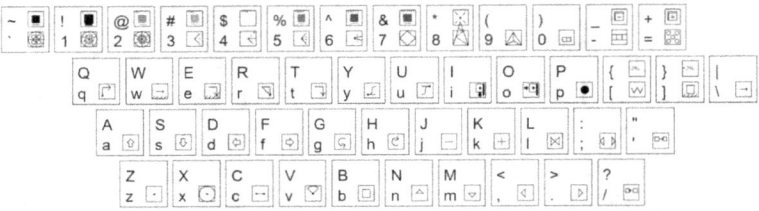

Figure 9.1: Keyboard layout. Strike a key to add an action to the glyph you are editing. Arrow keys and backspace allow for editing the whole glyph. Up and down arrows move to the end or beginning of the glyph.

menu bar at the top of the screen. When you have saved a symbol you can see it and download it from the Symbol Feed, which is an icon three in from the left showing equilateral triangles separated by an arrow. Each symbol is stored as both a bitmap in .png format and a vector graphics file in .svg format. What is listed in this Feed is simply a sequence of stored files in the directory symbolfeed/, which exists on every Geometron server. Each file is named with the UNIX timestamp which describes the exact moment it was created. The Symbol Feed, like all Feeds in the System, also has delete buttons to delete any of the symbols in the Feed, following the Law of Geometron that everything dies. If you click on an .svg file in the Feed, it will load that symbol into the editor, and you can then go back and edit a copy of it, and save it again to make a modification. The Symbol Feed also has an upload button just like the Local Image Feed, which lets you upload .svg files from other systems and allow you to edit them using the copy as a template. Thus we can freely replicate, edit, delete and replicate again.

Move the cursor around, draw circles and lines and squares. Make polygons, explore scales and symmetries. Try out the Bezier path drawing. Play with colors. Make closed and open paths. When in doubt, destroy it all and start over.

As you create and edit the main glyph, the address sequence will update live in the text field next to the one where the cursor is for editing. This will be a sequence of numbers all of which begin with 0. As with all other parts of the Geometron system, the ability to share information instantly across the world with it is of the highest importance. If you copy the sequence of addresses in that field to the clip board and send it via text message, email or pastebin to someone else with Geometron, they can paste it in that same field and hit return and they will instantly load the same glyph you created, which they can then edit and send back to you modified.

What you are editing in the system is not just a glyph but a whole JSON structure which is how we can share a whole symbol including style, position, scale, and custom symbolic language tools. This structure will be described in more detail below, but you can immediately start sharing them by clicking on the icon which says JSON in the

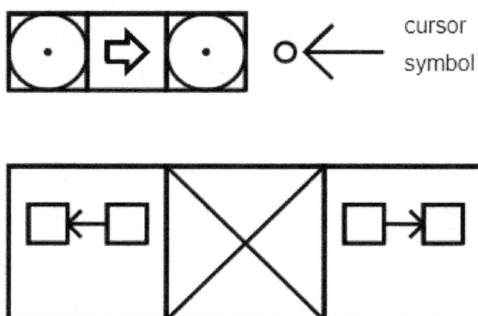

Figure 9.2: Edit symbols. This shows the symbols for moving the cursor back and forward, deleting an action, and also shows what the cursor looks like in a glyph being edited.

Symbol app. This will bring up a screen which has the JSON data in a text area, and a set of buttons to export, import, save, or reset. Reset is important because it is possible to corrupt the data beyond recognition and this gets you back to something which will definitely work. As with all other components of the Geometron system, this is a human readable text format which is designed to be shared both by direct text messaging and by copy and pasting into public pastebins and sharing the pastebin link.

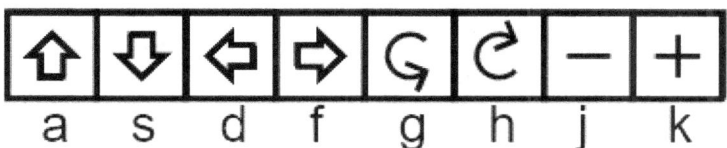

Figure 9.3: Movements. Arrows move along directions of the lines in the cursor. Rotation is by the unit indicated by the cursor wing angles. Scale actions are by the current scale value as shown by the dot positions on the cursor. Letters shown indicate the keys which map to these actions on a QWERTY keyboard with the default settings.

Each instance of the Geometron Virtual Machine has a style object, which defines 8 layers, numbered from 0 to 7. Each style has a line color, line width, and fill color. The properties of the style object are stored in the JSON file data/currentjson.txt which is used by the app symbol.html to edit graphics which are used by the rest of the Geometron system.

While the style app edits the data file currentjson.txt which applies to the whole Geometron object used for

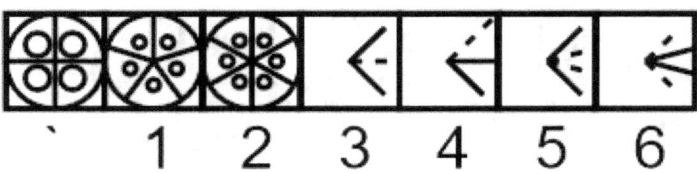

Figure 9.4: Angles described by symmetry glyphs. This also shows the actions to bisect, double, trisect and triple angles, and what keys are used to activate each geometric action.

symbol editing, the importing and exporting of data for sharing with other users only includes style information, without the rest of the JSON data. This allows styles to be separated from the rest of the information for the purposes as usual of building a robust remix culture where Geometron users can constantly be sharing each piece of the system. The EXPORT button will always post the current style JSON in the window in the lower left of the screen. IMPORT will import the data, and RESET returns it to a default state. Try creating your own new style with unusual line widths and colors, then exporting it and saving it offline, sharing it with other users, etc.

Colors are in the format of HTML/CSS/JavaScript,

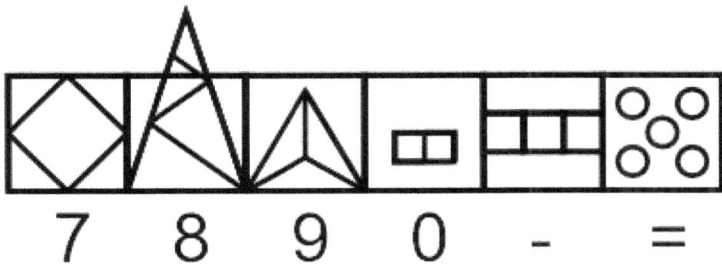

Figure 9.5: Scales, along with keys used to map to them in default configuration. There is no relation between the numbers on the keys and the mathematics of the scales. The scales shown are, from left to right, the square root of 2, the Golden Ratio, the square root of 3, 2, 3 and 5.

and can be either names of colors like "red" or RGB color values like "#00ff00". This last format is a number in base 16 which has three 2 digit numbers in it(numbers between 0 and 0xFF), where the three numbers are values of red, green, and then blue. So black is #000000 and white is #FFFFFF. Any value where all three numbers are the same, like #808080 will be a shade of grey. Colors can be partially transparent by adding a fourth hexidecimal number which represents opacity. So fully opaque red is #FF0000FF, and red with half transparency is #FF000080(80 because 8 is half of 16, this is actually

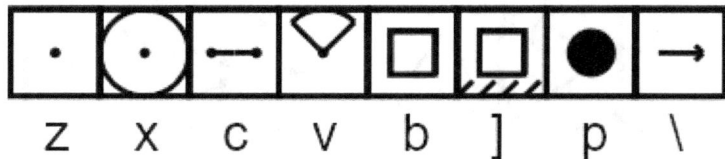

Figure 9.6: Basic drawing actions, along with keys used in default configuration to activate them. From left to right the actions are: draw dot, draw circle of unit radius, draw line segment of unit length, draw arc between cursor wings, draw a square, draw a filled square, draw a filled circle, and draw a line segment while moving forward one unit.

128 in decimal).

The next section of the JSON we want to know how to edit in order to be able to make useful graphics is the setup, edited in the app setup.html. Setup edits five numbers, all of which are in units of pixels: x0, y0, unit, width and height. Width and height are the width and height of the graphics file currently being edited or created. When a Geometron glyph is drawn with a given GVM, it starts with x and y equal to x0 and y0. Setting these two values is therefore effectively setting the

Figure 9.7: Layers. Each layer has a line color, line width, and fill color, all of which are set with the Style object using the Style editor app.

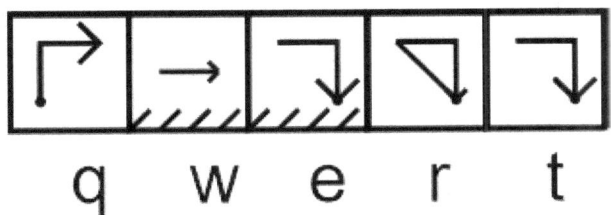

Figure 9.8: Path actions, with keys used to activate them in default state. From left to right, actions are: start path, draw line segment in path, close a filled path, close an unfilled path, and terminate a path without closing it.

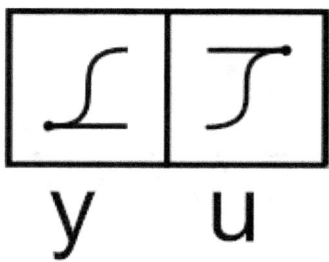

Figure 9.9: Start a Bezier Path and terminate it with the y and u keys.

horizontal and vertical offset of the field of view of the symbol. When we activate a pan function within the symbol.html app what we are really doing is modifying the values of x0 and y0 in the JSON file. These are done manually in this app. Finally, unit describes the initial unit value of the GVM. This is essentially the scale factor. So again when we activate the zoom functions in any other symbol editor what we are really doing is making changes to the variable unit in the global JSON file.

The app setup.html has five fields in which to enter the numbers for the five values. There is also a reset button to restore default, with a 600 by 600 pixel square and 80 pixel unit centered in the center.

The final section of the global JSON file which defines

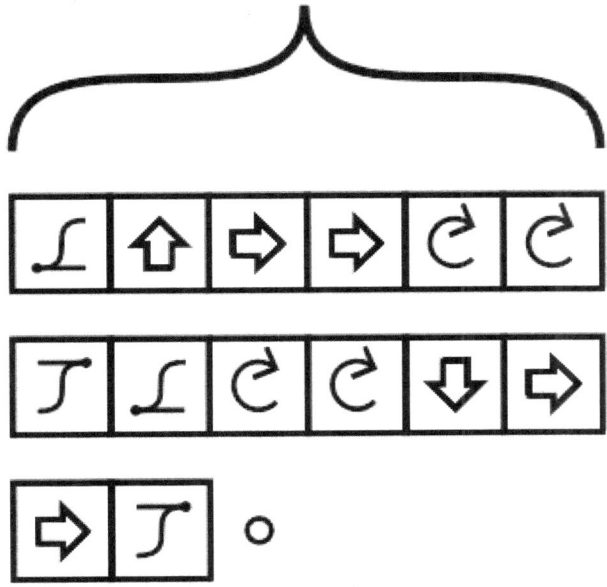

Figure 9.10: Demonstrating the power of Geometron to make useful symbols with Bezier paths quickly and easily: a twiddle bracket.

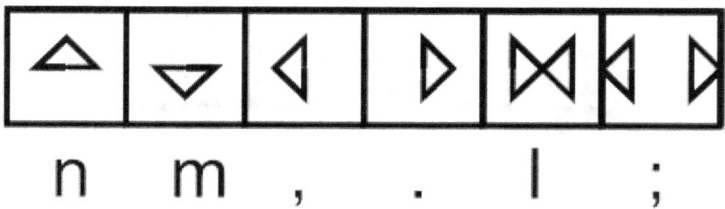

Figure 9.11: Pan and zoom the field of view.

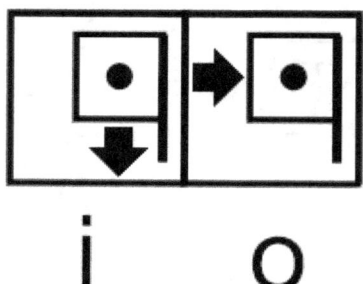

Figure 9.12: Drop a flag, return to flag. This saves and then recalls the state of the GVM.

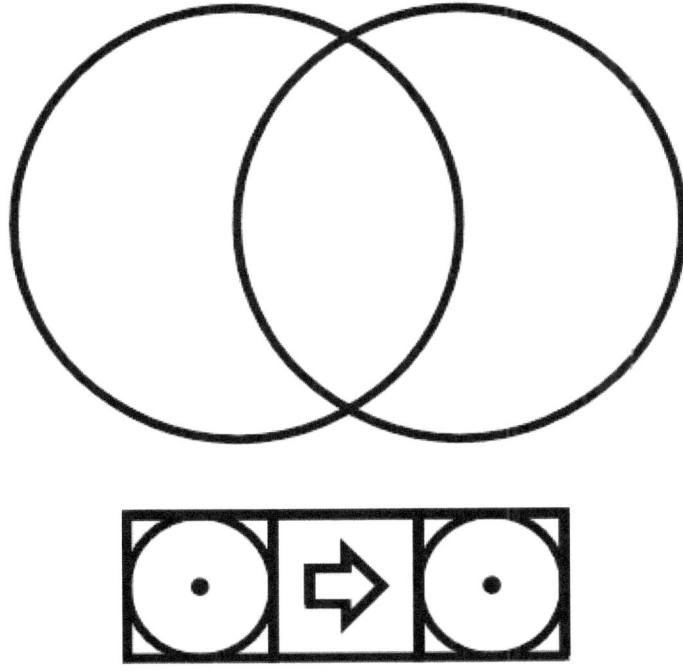

Figure 9.13: The "hello world" of geometric programming, the Vesica Piscis.

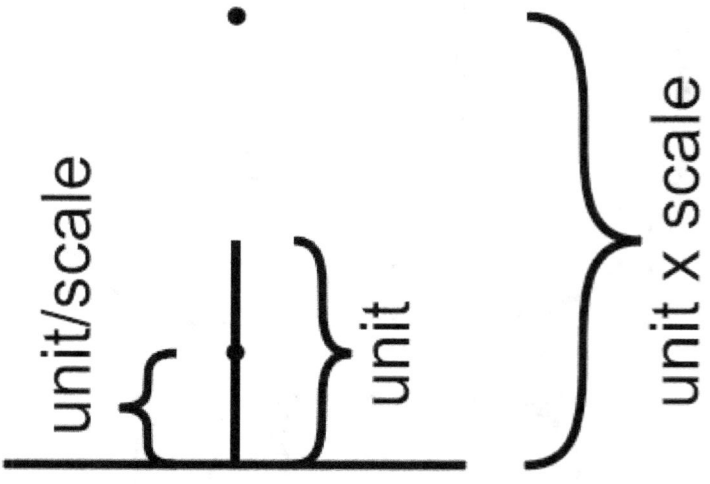

Figure 9.14: Cursor scale. This shows how scale works with the Geometron cursor.

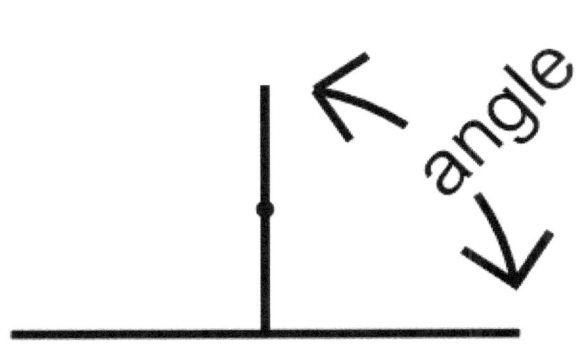

Figure 9.15: Cursor angle. This shows how angles work with the Geometron cursor.

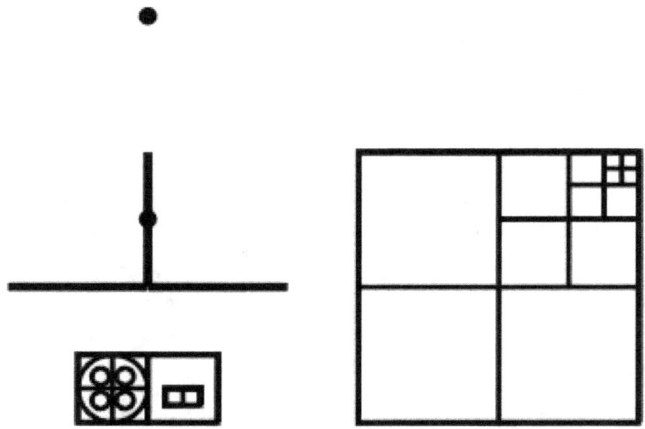

Figure 9.16: What the cursor looks like with factor of two scaling and a 90 degree angle. Also shown is a square used in Action Geometry. Try making the square!

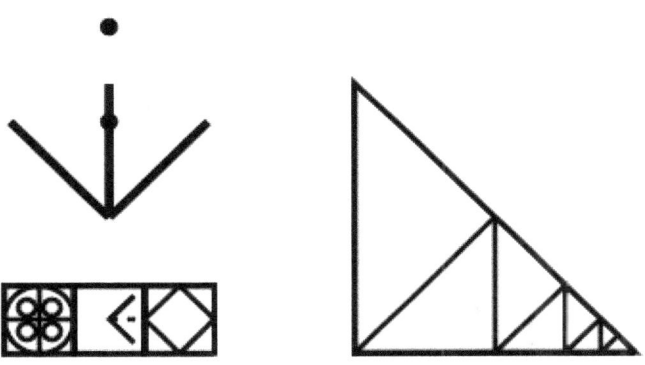

Figure 9.17: Another example of a commonly used Geometron cursor state, which combines the square root of two with 45 degree angles. Also shown is yet another shape used in Action Geometry which is also a good exercise to try to copy yourself.

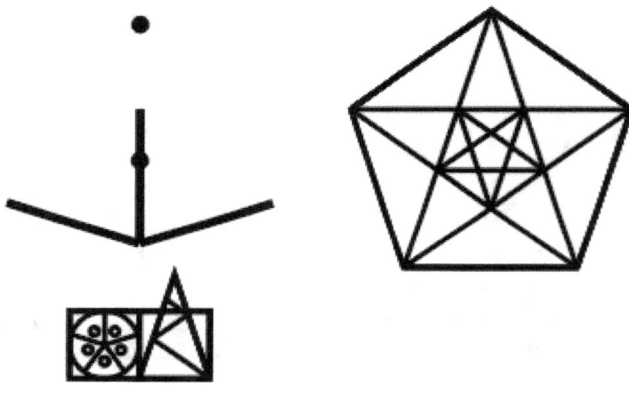

Figure 9.18: Cursor with Golden Ratio scaling and 72 degree angle for fivefold symmetry work. Shown is another shape that is helpful to learn to copy, the pentagon/pentagram fractal.

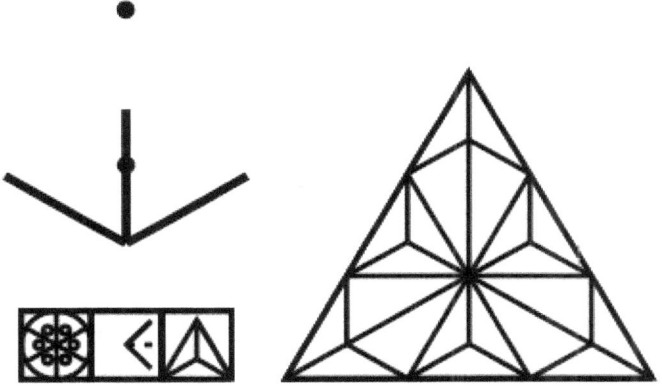

Figure 9.19: Cursor with square root of three scaling and 60 degree angle. This can be used to make the kinds of symbols shown, and replicating that is a useful exercise, as well as working through the deconstruction cf the six pointed star and hexagon.

154 CHAPTER 9. 2D WEB GRAPHICS

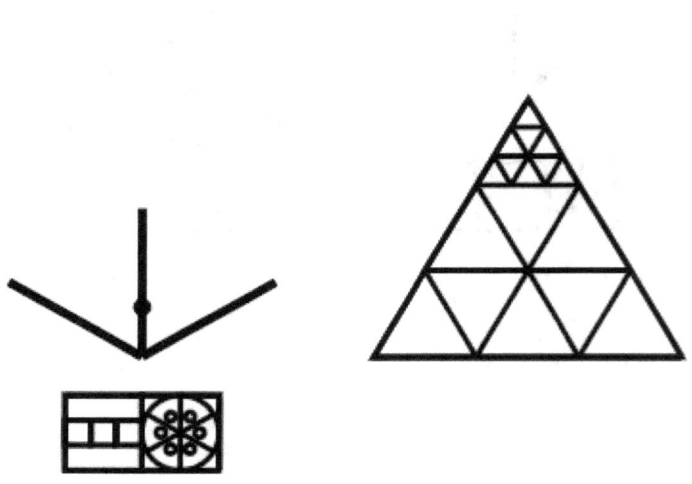

Figure 9.20: The cursor with a 60 degree angle and factor of 3 scaling, along with another exercise to copy.

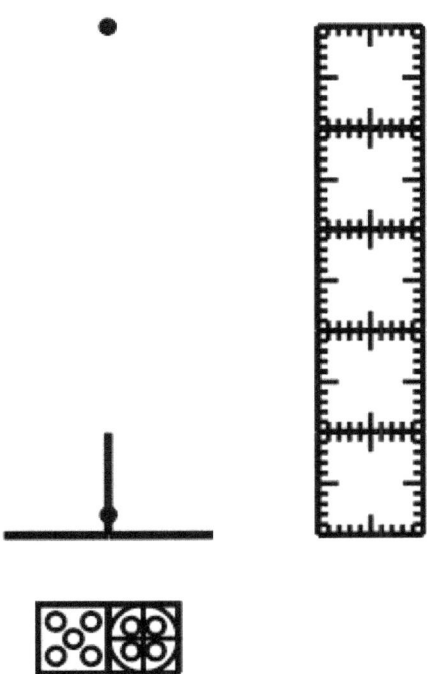

Figure 9.21: Cursor with scaling factor 5 and right angles. This can be used along with scale factor 2 to make things with scale factor 10. What is shown to copy as an exercise is a ruler constructed using this tool which can be made physical using a laser cutter as discussed in the Action Geometry chapter.

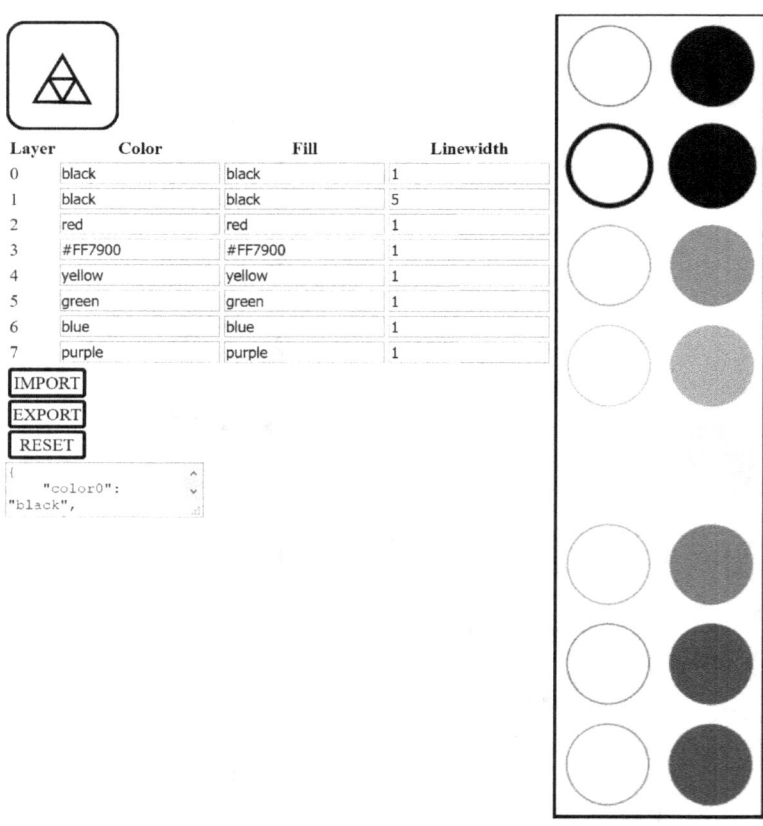

Figure 9.22: Screen shot of the style editor app at styleeditor.html. The display on the right hand side of the screen shows an unfilled circle and filled circle of each layer's style. The text area in the bottom left of the screen is used to import and export style data, which can be saved offline and shared with other users via text message, email, etc. The RESET button resets the style to a standard setting, which will erase any changes made to the existing style. Enter new values into any field to immediately change it.

the settings of the symbol app is the shape stack. This is a subset of the hypercube, and is stored both in the JSON file data/hypercube.txt and also data/currentjson.txt. When vector graphics files are saved, this shape stack is stored inside them so that they can be reloaded with the whole stack. The creating and sharing of useful shape stacks will be discussed in the next chapter. This is a very important element of the system as it allows for the rapid dissemination of specialized graphical languages for things like drawing circuit diagrams or subway maps.

Other links to other apps from the main symbol app are to the keyboard editor, which edits the layout of the keyboard, the Hypercube editor, which edits the entire Geometron Hypercube, and the Font editor, which edits just the font. All these have the capacity to share human readable text from system to system as is the case with everything in the Geometron system. The font and Hypercube apps are still somewhat crude and could be improved substantially, but they do work for their intended purpose with a little bit of fiddling. They will be used through the rest of this work as we delve into more applications of Geometron geometric programming.

There is also an app separate from the main symbol editor called symboltrace.html, which allows you to trace images into symbols, which you can then copy and paste at will. This takes images from both the local and global image feed, so you will need to load the image you want to copy in one of those first. Still another random app of some utility is action2symbol.html, which converts a glyph made up of actions to one made up of the symbols which correspond to those actions. This is how we put glyph symbol spelling into a symbol, which is very useful for documenting things which reference the language and how it is used.

Part of the power of the Symbols in Geometron is how they are integrated into the rest of the system. When you save a symbol, a copy of the base 64 encoded bitmap is stored in the textfeed which gets used by the Map editor to create maps. This means that you can create a symbol, then go import it into the Map editor immediately. Because it is a self-contained image url which has the actual graphics encoded in the url rather than as an external file, this Map you create can then be shared with anyone else anywhere in the world with another Geometron system and they will be able to import and use that Map without any other image files. This ability to instantly integrate Symbols into Maps can enable a sort of graphical meme system which can be extremely powerful for numerous applications.

Also, we note that all icons used in the Geometron system are created using the Symbol editor described in this section. These are then stored in the directory iconsymbols/. This directory is all copied with every copy of the whole self-replicating system. Therefore if you want to add another symbol to the next copy of the system, just add your new symbol to this directory, run dna-generator.php, and the next instance will have your new image.

If you are a coder, you can read the whole of the system documented in this chapter in the JavaScript library stored at jscode/geometron.js, which is on every instance of the system. You can of course edit this, and your edits will replicate along to the next instance of the system.

Chapter 10
Shapes and Fonts

To see the full power of the Geometron idea of geometric programming, we must explore how we build custom graphical languages. Language customization happens by editing the contents of the Geometron Hypercube, the geometric structure described in the Symbols chapter which defines the meanings of geometric actions carried out by the Geometron Virtual Machine. Each action also has a symbol. So to create a new graphical language, we decide what graphical elements we want to create, build those, and then also build symbols for each one which fit the format of fitting inside a square with a left to right progression so that they can be part of the glyph spelling we use when creating and editing glyphs.

This chapter is more technical than most people will need or want, and is here for completeness so that the system is fully documented. Primarily it is a gallery of pictorial examples, and the reader is encouraged on a first reading to skip the text and look at the pictures before moving on to the next section. This is partly intended as a stub of sorts, a documentation of software which people can re-write to be easier to use as the system evolves.

In principle, there are hundreds of addresses in the Geoemtron Hypercube which we can edit to create custom languages. These are divided up by function and will be dealt with in different sections of this book. In practice when we use the basic symbol editor for making 2d web graphics, by far the most commonly used is the Shape Stack. This is a selection of addresses which are reserved for building languages which have up to 16 shapes, with a matching set of 16 symbols. The important thing about the Shape Stack is that it gets saved inside each .svg file when they are created in a human readable format which can be extracted back out of the file either automatically or manually. This is how it is possible to click on a .svg file in the Symbol Feed and have the correct symbol load up and be edited, even when symbols have vastly different specific graphical languages they use. Engaging the 16 elements of the Shape Stack is done in the default keyboard configuration with capital letters in the second two rows on the keyboard from left to right, or Q, W, E, R, T, Y, U, I, A, S, D, F, G, H, J, K.

To edit the Shape Stack, we use the shape stack editor which is at shapestack.html. This is linked via an icon with a 4x4 grid on it. The shape stack editor uses only the keyboard, and does not have a touch screen interface at the moment. To edit a shape, you use the customized Geometron keyboard just as you would for creating and editing glyphs in the main symbol editor. To move to the next and previous shapes, use the up and down arrows. Try stepping through up and down and get a feel for that. Also, don't forget that you can use the pan and zoom actions to change your view to zoom in on details you are editing, and that these actions are located at the lowercase "l", the semicolon, and n, m and comma and period.

After you create a new shape, you can change the symbol which represents that shape by clicking on the hypercube icon to go to the shape symbol editor at shapestacksymbols.html. This is basically the same editor, but pointed at the addresses corresponding to the symbols of the actions you edited in the main editor. We always want to edit the symbol for a given action so that at the end the cursor is one unit to the right, a square has been drawn, and the state of the GVM is set back to 90 degree angles and factor of 2 scaling.

The shape stack editor also allows us to trace over images, and you can click on the images in a window which lists the images in the image feeds to select an image, then use the slider bars for scale and rotation to set the image where you need it to in order to trace it.

As with all elements of the Geometron system, the

most important thing we can do with a shape stack is to share it from person to person across the world. And as always, this information needs to be human readable text in the smallest possible format so that we can copy it to a clipboard and paste it in text messages, emails, directly from browser window to browser window or into pastebins which we can share. The format for a shape stack is an array of quoted text inside square brackets, separated by commas each of which has an address followed by a colon and then a sequence of addresses which represent the glyph stored at that address. Once one gains familiarity with the address system of the Geometron Hypercube it is possible to read this code and manipulate it by hand. However, for people who don't want to dive deeply into any of this, or even to learn how to edit it or interact with it at all, the important thing to know is just that you can use the export and import buttons to exchange shape stacks between one Geometron server and another.

Also, if you want to see the shape stack used in the creation of a Geometron symbol in .svg format, you can open that symbol in a web browser and use the "view source" feature which exists in all browsers. Inside the source code for the .svg you will see a little bit of JSON code at the top inside an XML comment, as well as commented out XML tags which mark where the json is. This JSON can then be manually copy and pasted into the JSON importer of the whole symbol system or just the shape stack can be removed and used.

If we want to edit and share Hypercubes, we use the Hypercube Editor, also linked from the main symbol app at symbol.html. This is at hypercube.html on every Geometron server. In this editor we can edit all parts of the hypercube which contain Geometron glyphs in the byte code format which is sequences of base 8 addresses. These include all the symbol glyphs, as well as the 8x8 tablets stored in the address ranges starting with 2, 5, and 6. Once again, we can export and import using buttons and a text area as in all parts of the system. The actual file being edited with the Hypercube editor is stored at data/hypercube.txt, and you can always look at that in raw plain text and copy it from there as well. Again, this editor is a sort of stub, and as our swarm grows, more and more people will create their own editors for the Hypercube and various parts of it, as the logic is very simple and it is not hard to make a much better editor than the one presented here. Also, in future editors, the fully three and four dimensional structure can be brought out, using the three dimensional web graphics described in a later chapter.

The font editor works the same way as the Hypercube editor, with the same basic edit functions but for the range of the Hypercube which holds the font: from 01040 through 01176. Again, this is designed to be shared, with text based import and export of the human readable text which holds the byte code. The two most important fonts we use are the Robot Font which has pixels which are drawn by the various robots described later, and the Laser Font which has cutouts which when used with a laser cutter can make stencils for spray painting text onto physical objects. This is part of how we jump the gap from the digital to the physical: text describes a font, text describes a glyph, then that exports to a vector file, and that loads into a laser cutter which makes a stencil which makes a physical link in a physical space which points to a domain which points to the physical location of a server which contains all the information required to replicate that entire system.

    The rest of this chapter is simply a gallery of examples of using the shape stack and fonts to make and use different graphical languages to show the potential of Geometron. Also blank pages and margins can be used to add hand drawn graphics documenting graphical languages for custom illuminated manuscripts to exchange for barter via the Street Network.

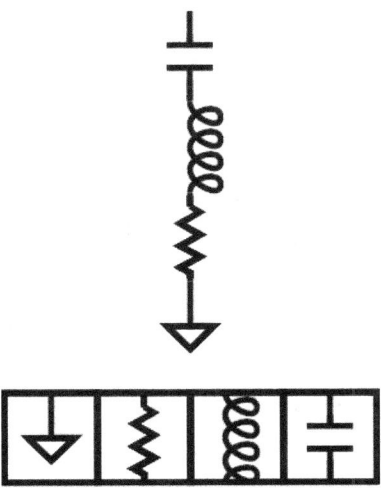

Figure 10.1: RLC circuit. Circuit diagrams are a great example of how powerful it is to create a custom graphical language. Once the schematic symbols for the components are programmed into the Shape Stack and mapped to keys on the keyboard, it is possible to create any circuit diagram with just a few keystrokes. This simplicity of construction also makes it possible to use any other software tool to generate schematic diagrams automatically. The symbol glyphs shown spell out the symbol, showing the extreme simplicity of construction once shapes are in place.

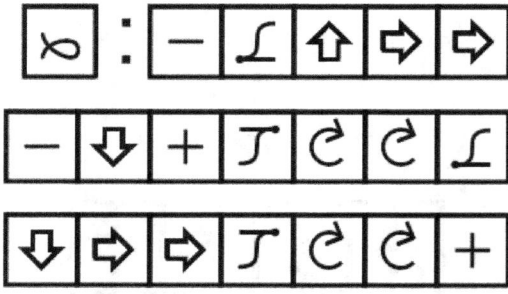

Figure 10.2: Construction of a single loop of an inductor. Once this loop has been created, it can be used to make inductors of any length as well as transformers and any other schematic symbol of something with inductive elements. This shows the power of building up symbols as fractals. Note the geometric precision and simplicity which are not present in the commercial vector graphics packages.

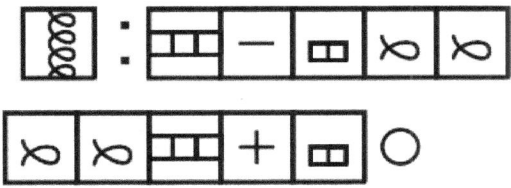

Figure 10.3: With the loop in place, just a few keystrokes makes a full inductor.

172   CHAPTER 10. SHAPES AND FONTS

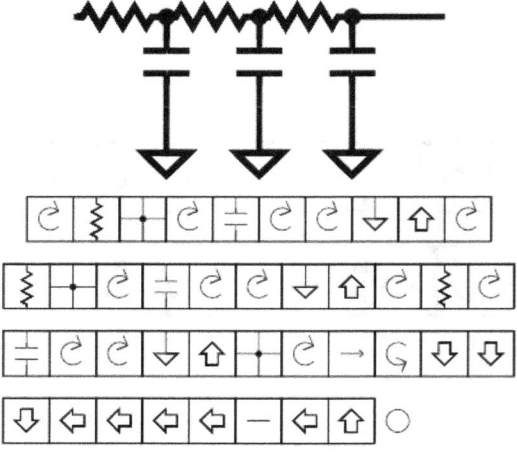

Figure 10.4: RC line circuit. Another demonstration of the power of Geometron for circuit building.

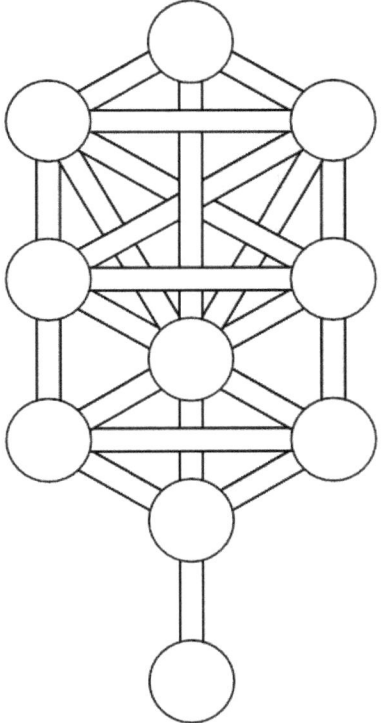

Figure 10.5: The Tree of Life from Jewish mysticism. This symbol also is used in various occult systems. There is quite a bit of geometry hiding in here, and you can see numerous published examples which get that geometry wrong, breaking the six-fold symmetry of the hexagon. With Geometron we can build a language for moving around on a hexagon and creating all the various bridges between nodes.

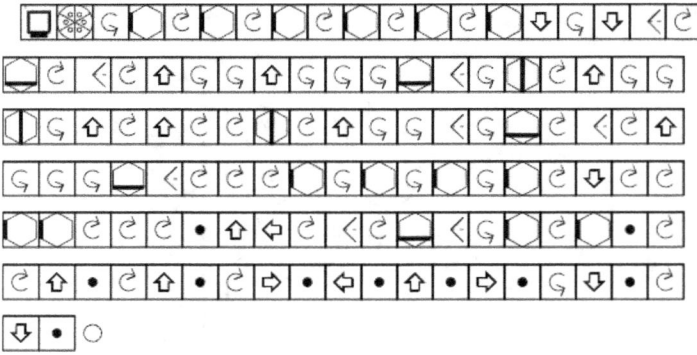

Figure 10.6: Symbol glyph spelling of Tree of Life. What makes things like this easy to make is building up building blocks like the cross pieces of all different scales, and the use of universal symmetries and scales(6 fold, 12 fold, and the square root of 3 and 2).

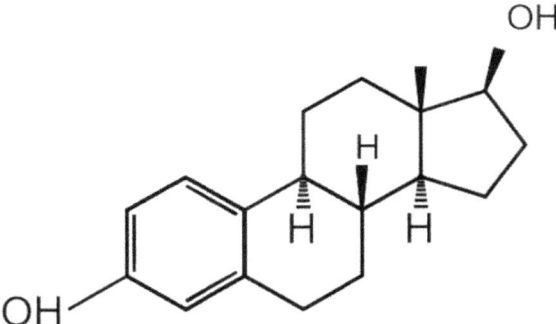

Figure 10.7: Molecular symbol for the steroid hormone estrogen. Note the characteristic hexagon-pentagon combination which is repeated throughout self-replicating chemical systems(life) as well as throughout the Geometron system/language.

176   CHAPTER 10. SHAPES AND FONTS

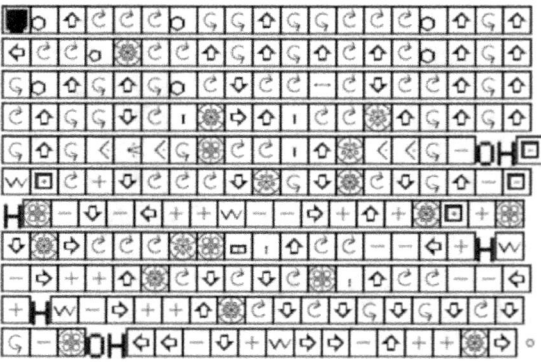

Figure 10.8: Symbolic spelling of the chemical symbol in the previous figure. This spelling is only possible because of constructing a symbolic language specific to drawing organic chemistry symbols.

Figure 10.9: Some letters from a Hebrew font. Typesetting a language which goes from right to left is a powerful demonstration of the way Geometron works. Each letter is a sequence of geometric actions, just as it is when a human uses a pen to create it. This means that if the actions end with the cursor to the left or bottom or diagonally that is where the next letter will be. There is no need to mode switch based on direction, the direction is built into each individual character.

178   CHAPTER 10. SHAPES AND FONTS

Figure 10.10: A katakana font using pixels. Again, we can create this with a sense of direction built in, which for Japanese has a few choices. We make this vertical so that it can be used to make Matrix rain, which is katakana in mirror writing. The font is made as regular characters, and then by simply changing the value in the hypercube of movement from left to right and from right to left from pixel to pixel, the katakana can be made mirrored or not without affecting the actual font elements.

Chapter 11
Action Geometry

We create a geometry in which constructions are whatever methods are most effective for replicating objects from trash and natural materials. In classical geometry many of us learned in school, we are restricted to the use of the compass and straight edge. Then in the geometry used in computers, everything is reduced to numbers and geometry becomes arithmetic. But in Action Geoemtry we use the technology available today to make shapes which carry information about the symmetries and scales we will use for common constructions.

Unlike in school where we learn to construct things at random, in Action Geometry the purpose of all geometric construction is to replicate things made out of trash in the way most beneficial to the people in our community. So to choose geometric tools we look at what is available and what we can build, and then build up our geometry around that.

The materials we start out using are cardboard, scrap cloth, bamboo, and high density polyethylene from milk bottles. We also use some commercial off the shelf parts and tools, as well as all the tools of Geometron. The Geometron language documented in the earlier chapters can be used to make geometric shapes which can be sent to a laser cutter. The shapes can also be traced off of a screen with thin paper and a pencil, the paper cut out and laminated, and then used as a physical construction. Or if you have a paper printer, you can print the file you create with Geometron. The shape set is shown in one of the figures. It includes a collection of shapes that all use 3 inches as a reference, and which contain information about the standard symmetries and scales from geometron. This includes 90 degree, 45 degree, 72 and 36 and 108 degree, 60 and 30 degrees, and the Golden Ratio, square root of two, and square root of three. Along with a standard Geometron ruler 6 inches in length, this set of shapes forms the basis of our whole construction method.

These shapes can be used to construct a wide range of shapes very quickly which can replicate, using the plastic parts to create a template from cardboard, which gets cut out and then used to trace out the same pattern again and again. This is self-replicating geometry! You can imagine a design, lay it out using standard symmetry and scales on any Geometron server, then use that design as a guide to hand construct that same shape once in cardboard, cut it out and copy it. Also, you can take the text data for the symbol you created on the Geometron server, put that in a public pastebin and share that with another person. That person can copy the data into their local Geometron server, edit the layout, and then use their copy of the shape set to make their copy of the cardboard template, then use a sharpie and box cutter to replicate again and again. Note also that this replication is cascading, like a critical nuclear reaction: each copied piece of cardboard is itself a template which can be used to trace out to cut out another copy in more fresh cardboard. So the Mathematics of this if there is a large swarm of people going into a very large feed of corrugated cardboard is that there can be exponential growth of the number of copies of a cardboard shape for a certain number of generations. This shows how scaling in a self-replicating economy can do things that are not possible in a consumption/production economy.

We use the simplest possible constructions to achieve any given task. The first things we build are the ArtBox, which is a self-replicating box of art supplies. The figures show the layout of how you can use just a 6 inch equilateral triangle repeated 10 times in the right layout to make an attractive art purse. The skin is wrapped in colored duct tape applied from the Tape Snake, also described in the figures. The box contains the shape set and ruler of Geometron described above and in the figures, box cutters, sharpie markers, scissors, googly eyes, and more clothesline. With this set of art supplies, all you need is this ArtBox fully loaded with the Tape Snake, and you can transform a feed of corrugated cardboard trash into an output feed of the same box which is then used to make more boxes and so on.

Note that this ArtBox is made from a combination of the octahedron and tetrahedron. Using these two is a tool we can repeat again and again for simple design of useful structures with the minimum of complexity to replicate. Each of these also contains the open brand of Trash Robot, which is rainbow and googly eyes Each box has googley eyes and a black duct tape mouth to make a face. This is a recognizable brand identity, but it belongs to no one. It is not property. So it can replicate freely.

Another basic construction from cardboard is the Golden Pyramid, which is shown in two of the Figures. This can be used as the enclosure for a very wide range of technologies, from battery packs to Raspberry Pi Geometron Servers, to lights, robot controls, or stereo equipment. It has a 6 inch square base, a 3 inch square top, and each side has the same angles as for the Golden Triangle(72 degrees).

Textile arts are created by finding black cloth from scraps and plain black clothes and sewing on geometric patterns with bright solid rainbow colors of felt or some similar material. Text are created by starting with a square and removing as little material as possible in as simple as possible layout to depict a letter. The Geometron Raspberry Pi servers are carried around in black cloth bags with a Raspberry Pi Penrose tile layout as shown in one of the figures and a 6 foot clothesline Trash Tie drawstring. There are two types of Trash Tie: the six foot clothesline terminated in duct taped ends, and the 18 inch nylon parachute cord section with burned ends. Small black cloth bags are sewn with small trash ties as draw strings, and these have symbols sewn on them representing the different types of clay icon token described in a later chapter. Large bags are also used to carry around the printer which makes the clay tokens which go in the bags. Some figures of the textile are left blank, these are filled in by hand in the illuminated manuscript in physical copies, and shared in person along with the physical textile arts and crafts.

Skeletron is a method of building structures of all kinds for supporting technology as well as building shelters and light industrial infrastructure. The Trash Pole is a roughly 6 foot long piece of bamboo with quarter inch holes drilled just in from each end as well as in the center of the pole, all wrapped in rainbow colored duct tape from end to end. These poles can be joined at the ends by tying the smaller Trash Tie(18 inch nylon parachute cord) through the holes and tying a square knot. With ends joined, we can construct again a wide range of three dimensional geometric structures using basic geometry of tetrahedra, octahedra, and small deviations from those. These constructions can then be sketched out by hand or via Geometron and shared freely across the Street Network, again forming self-replicating physical geometric construction. These poles can be used to build useful things, which induces passerby to donate duct tape and go scrounge for more bamboo, and join in the work of assembly, making a swarm just expand potentially very fast, as long as the rate of new contributors is maintained. This again stresses the importance of building a powerful open brand. If the extreme rainbowcore aesthetic can be used to get peoples attention, that will make something which spreads in an organic way. Not viral, as it is not just information in a fixed system, but organically, as a physical thing is being replicated. The bamboo could be substituted with any other straight object, like broomsticks, fenceposts, harder solid sticks, etc. This whole scheme can of course also be scaled up and down with any other scale and material of straight thin structural element.

This type of triangular replicated structure can be very strong and can build up fractal trusses to make huge complex structures both on land and in and on water. Skeletron forms the basis of a totally decentralized modular construction method. The sticks can form frames, and to build shelters from rain and wind, plastic bottles can be cut into strips and those stripes woven and laminated into giant plastic sheets which can be affixed to the frames. This can be combined with cardboard and paper trash and things like polystyrene trash to insulate the structure, to make robust livable shelters which can scale based on our Action Geometry. To become an artist in this form of construction, you will want to study deeply and practice extensively with all constructions involving tetrahedrons and octahedrons. Experiment, study, document, share, replicate.

Because we have a lot of things designed to be hung from Trash Poles using Trash Ties, we also want modular hooks to suspend things from the Trash Poles. These are the S-Hooks, which are made from a stack of 4 corrugated cardboard cutouts wrapped in rainbow duct tape, with googly eyes for branding. The construction for this is done using the 3 inch square shape from the basic Action Geometry shape set, and is shown in the figure. Again, each time an S is cloned, that clone is the template to make more S's, so this part can replicate with an exponential growth if it is fed into the right trash feed with the right group of people in a local Trash Robot swarm.

This chapter describes then a set of objects which we can share in a physical place by a physical Geometron Server. These can be shared, used, given away, bartered, sold, improved upon, and sent along to other places in the world to seed new swarms. A Geometron Trash Camp might include any subset of the things described here. Huge camps might have a whole village of livable shelters made from Skeletron with Trash Sheet, insulation, HVAC systems, power, industrial production machinery, telecom infrastructure, water purification, sanitation, and agriculture, and span a substantial area. A small camp might just be a Server on a Golden Pyramid, hanging from a dumpster at a truck stop via a Trash Tie, delivering documents over a hotspot on a smart phone. But whatever tools and materials we use, we can rely on the power of the Geometron geometric language to replicate constructions from community to community via the Street Network so that we can reach all of Humanity who wants to share.

192   CHAPTER 11.   ACTION GEOMETRY

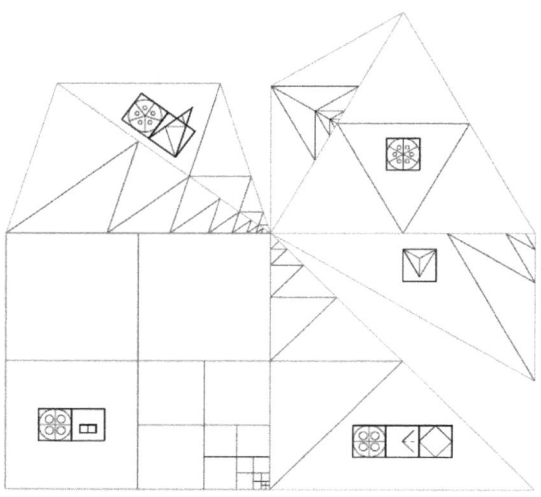

Figure 11.1: Shape Set. This is the basic shape set of Action Geometry. It has the symmetries and scales of Geometron. What is shown should be printed exactly 6 inches wide, making each shape three inches on a side.

Figure 11.2: Rulers. Make this 6 inches wide and each ruler is a 6 inch ruler, 1 inch across, with both tenth and factor of two divisions.

194   CHAPTER 11. ACTION GEOMETRY

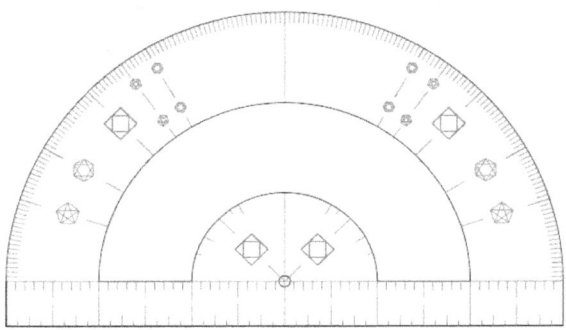

Figure 11.3: Geometron protractor. While not really needed for Action Geometry, this protractor is a nice accessory which emphasizes Geometron symmetries rather than numbers, and allows drawing of circles of radius 3,2 and 1 inch without a compass. This is mostly useful if cut out with a laser cutter.

195

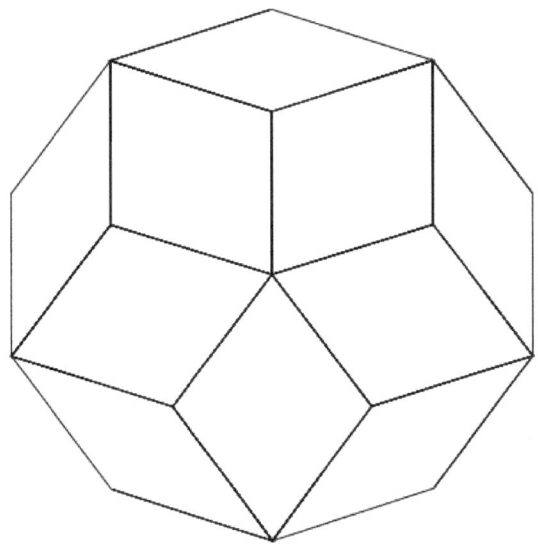

Figure 11.4: Penrose. Penrose tiles, the rhombi construction. Copy, print, trace, or laser cut these, and use them to make logos, icons and symbols with some kind of meaning which other people can easily replicate. The Shape Stack can by copied from someone which has these two shapes, and that can be used to create designs on Servers, which can then also be shared with your community who can all copy your highly recognizable design which has the Trash Robot metabrand as well as whatever symbol you have created or edited.

Figure 11.5: Spray paint stencil for laser cutting. You can use the Geometron software, selecting the built in laser font, to make a custom spray paint stencil pointing to the domain name which points to your Geometron server. If you are Trash Robot, that can be Trash Robot, but we mostly point to a local non-property place.

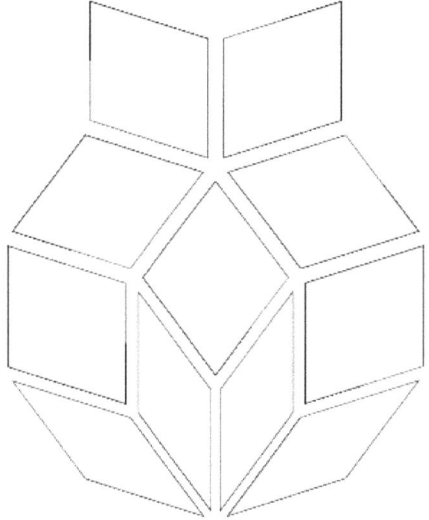

Figure 11.6: Construction of the Raspberry Pi logo for server bags. The top two shapes are green the rest are red.

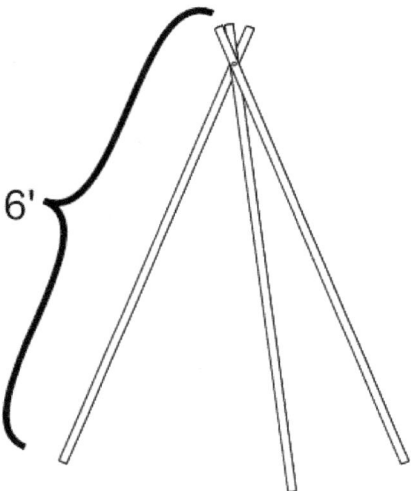

Figure 11.7: Skeletron tripod. Three 6 foot bamboo trash poles wrapped in rainbow colored duct tape with quarter inch holes drilled just back from the end. An 18 inch nylon parachute cord trash tie is used with a square knot to secure the top. Many things can be hung from this, including servers, terminals, robots, boxes, flags, lights, textile arts and pendants on more trash ties. The tripod can be carried over the shoulder to be mobile, without untying the joint at the top for rapid deployment.

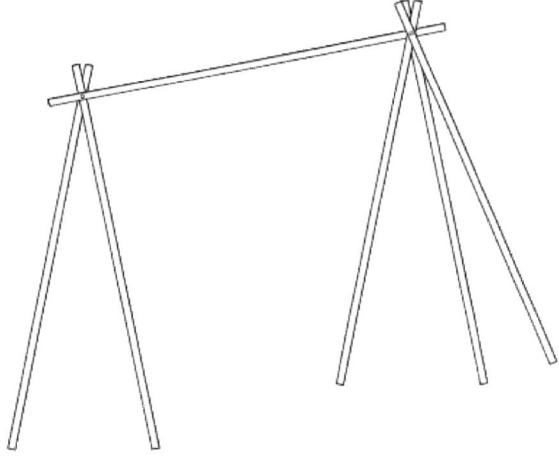

Figure 11.8: Skeletron cross bar configuration. Two tripods can be converted into this stable configuration quickly to have a horizontal cross bar which S-Hooks can hang from to hang numerous objects of all kinds.

200   CHAPTER 11. ACTION GEOMETRY

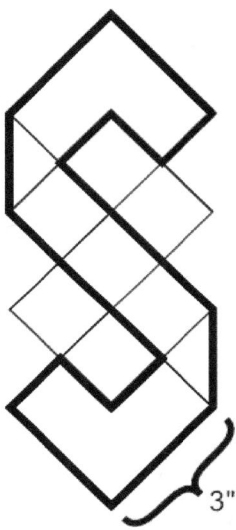

Figure 11.9: S-Hook. Note the repeated use of the square shape, and the use of the 45 degree triangle, making this easy to replicate using the Shape Set. This hook is used to hang things from the Trash Poles in Skeletron. Fabricate by stacking 4 identical layers of corrugated cardboard cut in this pattern and wrapping them in rainbow colored duct tape. Googly eyes can then be applied.

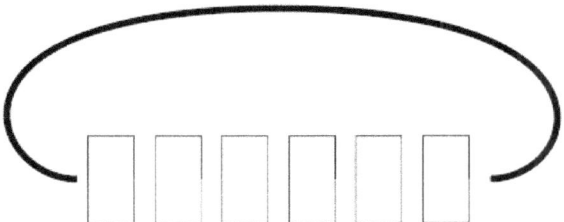

Figure 11.10: Tape Snake. Duct tape rolls of all useful colors, namely the rainbow colors plus black and pink, are strung on a 6 foot Trash Tie made from a clothesline which is looped through twice and secured with a square knot in a bight(like a bow on a tied shoe) for rapid replacement of rolls as they are used.

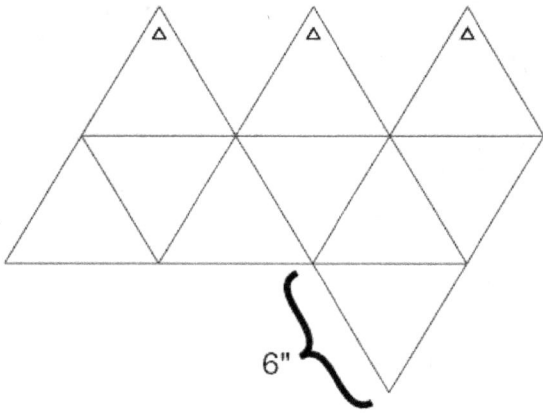

Figure 11.11: ArtBox Net. Cut out 10 equilateral triangles from corrugated cardboard. Duct tape the joints as shown.

203

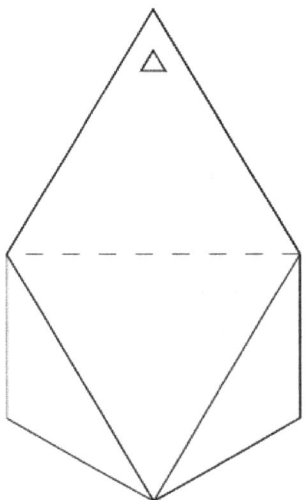

Figure 11.12: ArtBox assembly. The fully assembled box is a tetrahedron on an octahedron. It should contain the means of its own replication, which is a box cutter, a ruler, an equilateral triangle, and a sharpie, along with a Tape Snake for duct tape fabrication, extra Trash Ties, and googly eyes. Use duct tape colors in sequence to create a fully rainbowed effect, then apply googly eyes and add a black duct tape mouth.

204   CHAPTER 11. ACTION GEOMETRY

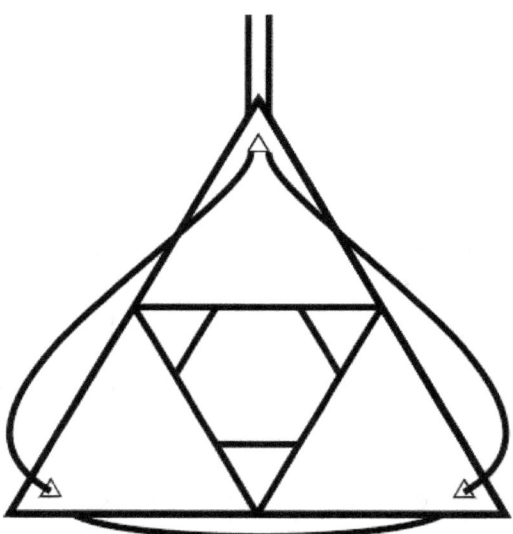

Figure 11.13: Top view of ArtBox. Cut out little triangles in each of the top three petal triangles with the box cutter. Thread a 6 foot Trash Tie with ends taped with duct tape as shown, and tie off the two bitter ends with a double figure eight knot for convenient purse-strap geometry.

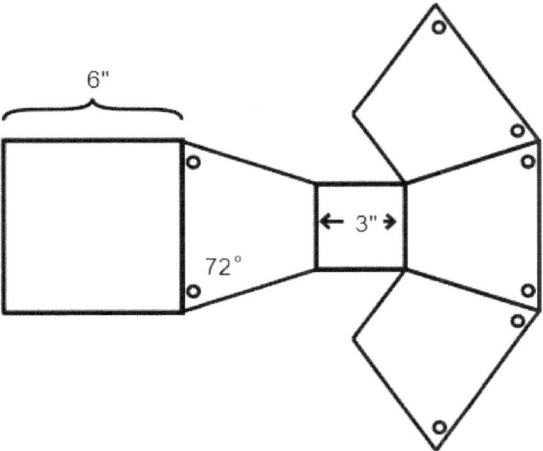

Figure 11.14: Pyramid net. Use the Shape Set and Ruler to cut out corrugated cardboard patterns and stitch together with duct tape as shown. Cutouts include a 6 inch square, a 3 inch square, and four trapezoids with 3 inch top and 6 inch base with 72 degree angles on the bottom angles.

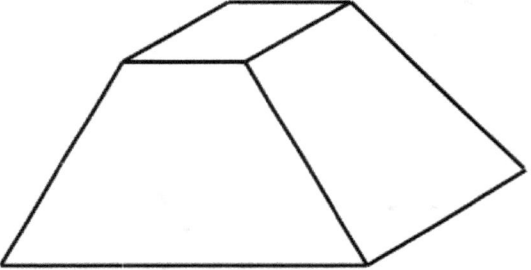

Figure 11.15: Pyramid assembly. Fold it all up and then cover the whole thing in a skin of rainbow duct tape. Open the base, insert technology and re-seal. Add cutouts as needed for cords in and out.

Figure 11.16: Bags are cut from black cloth, which can be scrap. Black cloth bags are sewn up with Trash Ties as draw strings.

Figure 11.17: Draw your personal Geometron outfit which is black with solid rainbow color felt or similar cloth sewn on in geometric patterns and geometric font.

Figure 11.18: Draw your Flag. A Flag is a black square cloth about 3 feet on a side, with a sewn hem with loops to tie a trash tie. It is decorated with solid block letters and geometric shapes cut from solid color rainbow felt or similar. All elements can be from scrap. Flags point to domains in places.

Chapter 12

Printers

In Geometron we view symbols as being any geometry with meaning, including physical constructions and sequences of actions. A "printer" refers to any machine which prints a symbol in this generalized sense. This means printer now refers to a huge class of machines, really any machine with automation can be thought of as a generalized printer of symbols. When we create symbols with the Geometron language, using the Geometron Hypercube and the Geometron Virtual Machine, we are building sequences of geometric actions, which can call other actions to build up complex fractal constructions. One of the kinds of actions we can build into this are discrete movements of machines, just as we use discrete movements of a virtual machine when the cursor moves around on the screen building two dimensional graphics

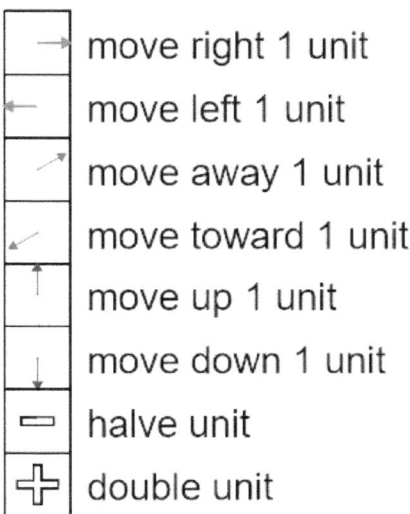

Figure 12.1: Basic geometric actions of machine control for an arbitrary machine that moves along three perpendicular axes.

in a web browser.

To build up control of machines for printers, we start with the most basic discrete movements. To begin with, we discuss the control of machines that have three axes of motion, all of which are perpendicular, along the up, down, left, right, forward, and back directions. We start with eight basic actions: a single step in each of the six directions, doubling the step size and halving the step size. Just these eight actions are enough to position a machine with three axis in any location in the available space. Just as binary numbers can be used to express any decimal number, this binary approach to geometry can be used to express any geometric location. Also note that these actions are independent of scale, and have no numerical units. A sequence of actions is all relative to some unspecified starting unit. So a glyph created to on some agricultural tool at the 100 meter scale can print at the nanometer scale using an atomic probe of some kind with no modification to the glyph in principle. Our language is both independent of numbers and of what machine is carrying out the actions.

Recall that each action has a symbol which can be displayed in a web browser. These symbols serve to allow us to edit glyphs which control machine action. Glyphs like this can be used to create the shapes in the range between addresses 0500 and 0577 in the Action Cube. These also have symbols in the Symbol Cube in addresses between 01500 and 01577. Whole universes of complex symbols can be built in the machine action shape table. Also, the overall system can have many more types of control than documented here if there are more components with more degrees of freedom for more complex machines, again with a full 64 potential geometric actions in the other action table from address 0400 through 0477.

The main initial application of all this we will dis-

213

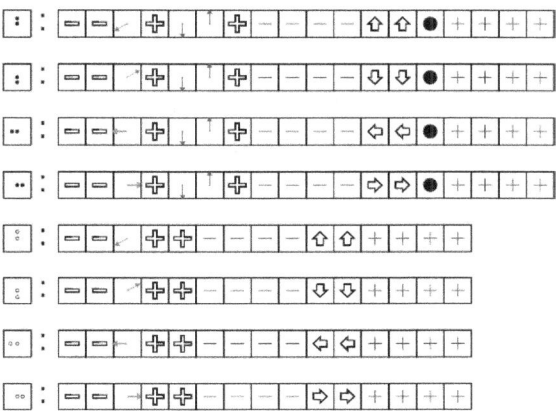

Figure 12.2: Dot actions from which symbols are constructed.

cuss here is printing out icons which are easily created and shared on the Geometron system. This method for printing symbols in physical objects can be a powerful tool for communication with other people about any idea about which people are capable of thinking or communicating. This printing system is based on a "two and a half d printer" model, in which we create three dimensional media on a two dimensional surface using a three dimensional tool to represent a two dimensional icon symbol. We are using the word icon here in a very general sense, just as it is used on layout of software systems: it is a symbol which can be used to represent anything, much like a word. Icons can in fact just be words, as there is a font built into the system. We build icons on the basis of four basic actions: move up, move down, move left, move right, move up with pixel draw, down with pixel, left with pixel and right with pixel. When a subject is chosen, we come up with a symbol for it, find that somewhere on the Web or draw it and photograph it and upload it, then align it and trace it with tools built into the Geometron system. Icon glyphs are stored in the Icon Feed, which can be used to share text by copy and paste with anyone anywhere.

We built printers from a set of 3 DVD drives, some cardboard and plastic, an Arduino and some custom electronics. When a pixel is drawn, a tool is moved down and back up in the z direction, and if this is done with a pointy tool over clay, it can create a dimple in the surface. The printer replicated in the Scrolls included with the system is about 14 by 5 inches, with a controller, as shown in one of the figures. It is for printing in round bits of polymer clay between 1 and 2 inches in diameter. Once the clay has been printed in, it can be baked in a home oven to harden it. Because it has dimples in it for each pixel, another piece of clay molded into the original print will make a mirror copy of the icon. When this clay is baked, it will form a stamp, which when applied to yet a third piece of clay will make a copy of the original print. This can then be painted and sanded flat so that the pixels are now colored in. Or it can be used to

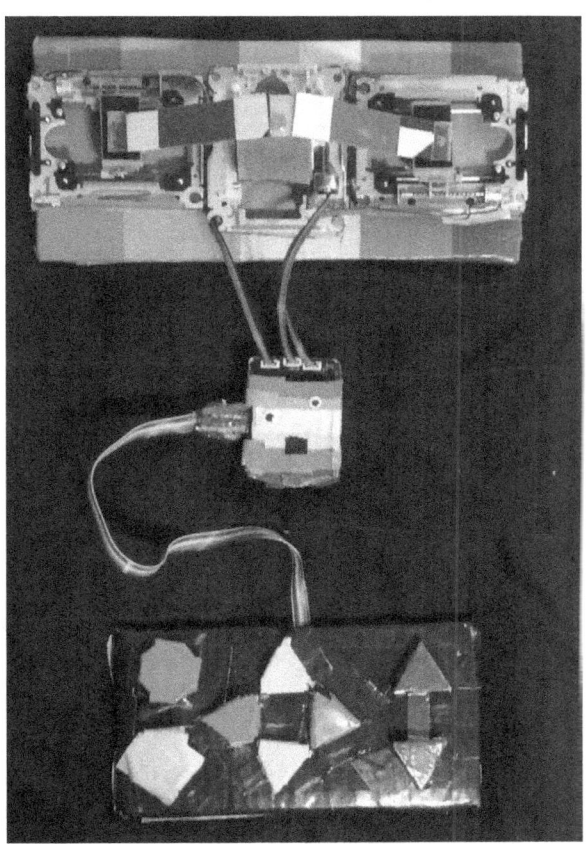

Figure 12.3: Clay Icon Printer. Printer is built from 3 DVD drives, cardboard and plastic trash, duct tape, an Arduino, and some custom electronics.

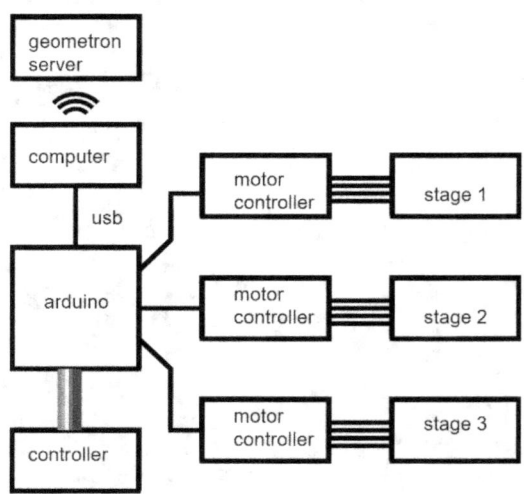

Figure 12.4: Block diagram of Trash Robot printer. The motion stages are salvaged from DVD or CD ROM drives. The motor drivers are off the shelf from Pololu Robotics. The Arduino is the Arduino UNO, and the whole system including the motors gets power from the USB connected to a computer. The computer can be used to interact with a Geoemtron server over the local network or on the global Internet to create programs which are copy/pasted into the Arduino IDE as described in the Trash Robot chapter.

print yet another stamp. Because an original can go to many stamps, the stamps can go to many more copies, and each copy can make many more stamps, we again have a geometric construction which can scale exponentially if there are number of people willing and able to carry out the copying action. Icons can also be used to make .stl files to send to a 3d printer, and .svg files with smaller holes which can be sent to a laser cutter for spray paint stencils. Since the 3d printed files also have pixels which can be pressed into an object to make molds, this is again self-replicating symbolic media. And of course with a spray paint stencil and a can of spray paint symbol replication can be extremely rapid.

The Icon Printer framework provides the basis for a very general way to represent any idea with self-replicating physical geometry. We will discuss in a future chapter how this can be a powerful tool for replication in the most general possible sense. With the abstraction of the 8 basic pixel movements fixed, this can also be realized in numerous systems besides the ones we have already shown which are included with the Scrolls. Two such potential versions are to go way up in scale and way down in scale. At the scale of a whole building, we can use trash-scavenged motors to build a system by which a little cart is pulled side to side along the edge of the roof of the building, which is controlled by one motor controller, and then a rope hoist which controls a winch which pulls a tool up and down along the face of the

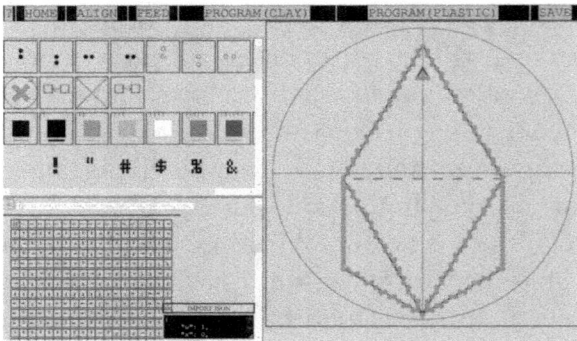

Figure 12.5: The icon tracer. Images are put in the Image Feeds as used in the rest of the system, then aligned in the icon aligner, and traced. When it is saved it goes into the Icon Feed for sharing with the world.

building. A can of spray paint is then added to the tool, with a simple mechanism to either engage it or not to draw a paint pixel. Once the motor controllers are set up to move along the basic 4 directions with or without a pixel, an Icon glyph can be loaded into the controller and an icon can be sprayed on the wall of a building at the scale of meters or 10's of meters. This might work better on the side of a dumpster or garage door. Scaling way down can be done either with electron beam lithography or optical lithography using the fine motions on the DVD drive stages. Lithography is carried out on polished brass, etching away metal everywhere but the

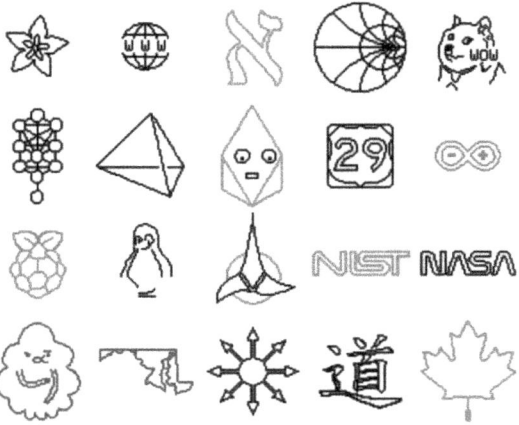

Figure 12.6: Icon Feed. Icons are just little bits of text, strings of numbers which are addresses in the Hypercube. Clicking an icon loads a glyph, which can be copy and pasted. New glyphs can be put in the text area and

pixels using a negative resist. Some kind of thick material like silicone can then be spun on the brass and then peeled off, making a copy of the icon. That process can be repeated many times, making in principle an endless feed of copies from the one brass original print. If this is in a clear material, it can then be put on a surface face down and the patterns will be visible. This can be used to print text, which can then be read with a scanning optical microscope connected to a cell phone camera, where

220   CHAPTER 12. PRINTERS

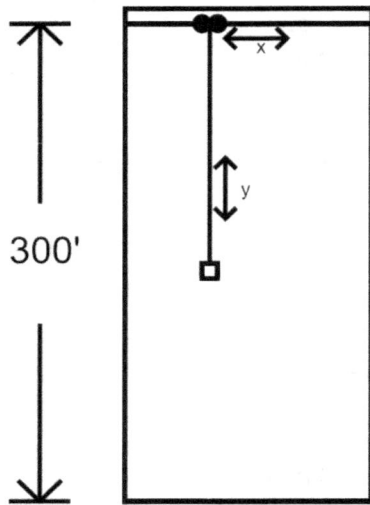

Figure 12.7: A hoist run along a rail going across the edge of a roof of a building can make a simple robot which can move to anywhere along the wall.

the mechanical stage for that is also built from trash.

In all cases, we use the Arduino as the basis for the controls. This versatile free and open platform can be controlled from the Raspberry Pi, meaning that we can have the full automation and printing system along with the server all be self contained without using any private machines. Thus the printers of self-replicating media can

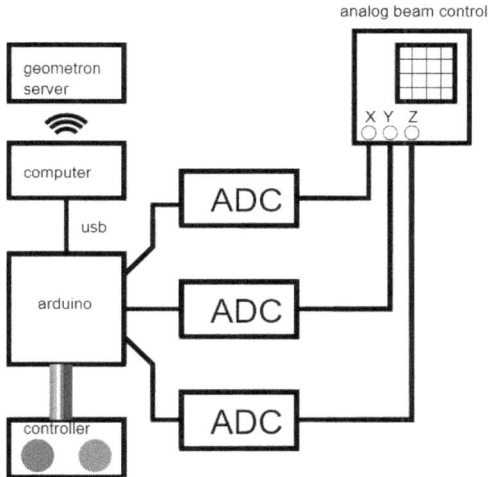

Figure 12.8: Block diagram of electron beam lithography Geometron robot. The Arduino here drives three analog to digital converters. Again the user can design a program for the beam path in a web browser using any computer, which can then also control and power the Arduino and its accessories. In this case the controller only needs one huge green go button and one huge red stop button.

be freely distributed over the Street Network, and anyone can create these and replicate them anywhere, and then share with anyone anywhere else in the world by a simple text message! This is truly a generalized information network. Any abstract concept can be reduced to a symbol icon glyph. That glyph can be instantly shared via text message with any of billions of people, and if they have a copy of our self-replicating server with our self-replicating software and our self-replicating robot built from trash, they can make a copy of the icon which is then itself self-replicating physical media. That can be used by them to communicate any concept in the most abstract possible sense to anyone else. They can then edit and remix the symbol indefinitely, and share it along with others and so forth.

The prints, stamps, and colorized icons described above are all carried around in black cloth bags. In addition, the stamps can be used to make pendants with holes at both ends that can be threaded onto Trash Ties from Action Geometry, making another attractive and independent form of personal wearable media which can be integrated into our system.

223

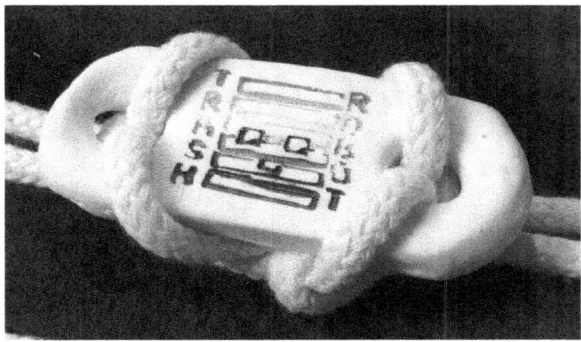

Figure 12.9: Pendant. One print can make many stamps, one stamp can make many of these. The back could also be unpainted for easier replication, making the object itself fully self-replicating media.

Chapter 13

Geometron in 3d and Beyond

As discussed in Symbols chapter, there is a layer of the Geometron Hypercube devoted to 3d geometric actions in software(as opposed to machine control). These actions are assigned the addresses between 0700 and 0777(base 8). For each action, there is a two-dimensional Geometron symbol with addresses equal to the actions, but with a 1 before them, ranging from 01700 through 01777(base 8). As with the rest of Geometron, we can edit a glyph which is made up of a sequence of geometric actions using an interface based on manipulating the symbols which represent those actions in a Web browser.

While it is possible to use the full grid of all 64 actions to create extremely complex three dimensional constructions, for the purpose of this work we restrict ourselves to a few basic actions. As with all other parts of Geometron we use the fact that discrete motions along the available geometric directions combined with halving and doubling of the unit of movement is sufficient to move a cursor to any place in space. We therefore have as our basic actions 6 discrete movements along three axes, the halving and doubling of the unit of motion, and the construction of a sphere and a cube, along with actions to select colors.

The addresses in the Hypercube from 0600 through 0677 represent "shapes" in three dimensional Geometron, which also have two dimensional Geometron symbols in the range from 01600 to 01677. These can be used to build up arbitrarily complex objects with layers of structural hierarchy. As an example of how useful this is, we will show how we can build up a universal rectangular base on which to print symbols.

In Geometron we are always concerned with making useful files which can lead to either physical production or have some function in operating the overall system. We also want to be able to replicate symbols created in other parts of the system in the three dimensional spaces we create using the software. To see what is useful for this, we have to discuss some of the formats and tools available for three dimensional graphics in Web browsers and in everyday use.

The most basic way to render three dimensional graph-

CHAPTER 13. GEOMETRON IN 3D AND BEYOND

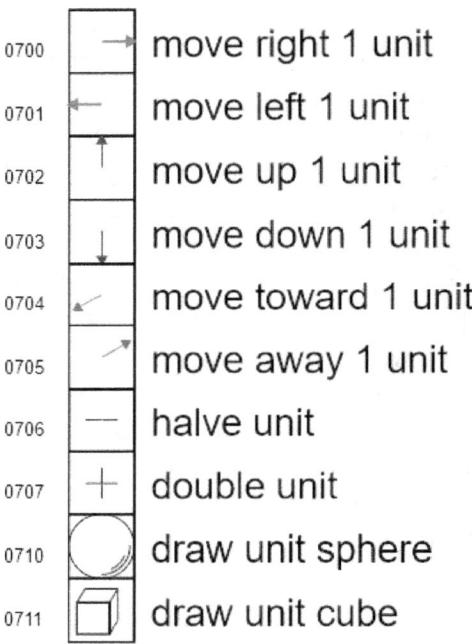

Figure 13.1: Actions. This is a minimalistic action set. More complex actions are added based on specialized applications.

227

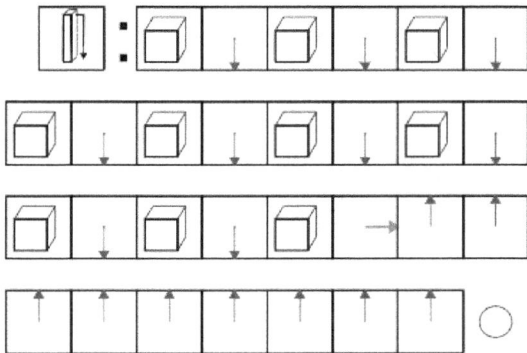

Figure 13.2: A linear rectangular solid can be built up using a sequence of repeated cubes. This 3d shape is then ready for use in more complex structures.

ics, and the oldest on the Web is to use the current updated version of what used to be called VRML, or Virtual Reality Markup Language. This was the original format back in the 1990's when people were first talking about a new way to exist on the Internet with ubiquitous virtual reality. This new web never materialized. Also, VRML was replaced by a newer more widely used format, called x3d, which is a subset of XML, just like HTML.

The most general three dimensional Geometron app

CHAPTER 13. GEOMETRON IN 3D AND BEYOND

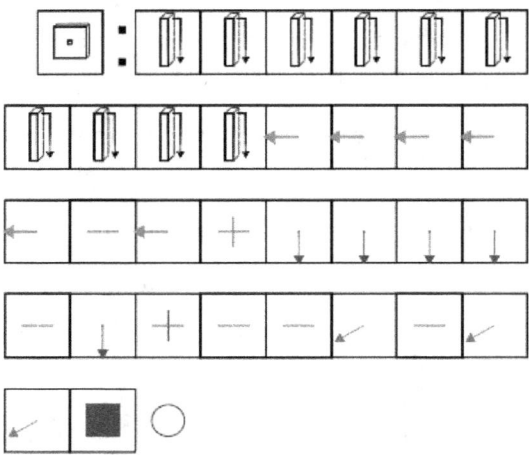

Figure 13.3: Linear rectangular solids can be used to build up a flat rectangular plate shape. The shape shown here builds this structure, then shrinks down to the size of a single pixel which will be used in the construction of 3d printed icons with the same bytecode as used in the rest of the system.

is called voxel.html. A "voxel" is the three dimensional extension of the idea of a "pixel", and this software can be used to quickly draw three dimensional objects in the browser using either softkeys or a special keyboard mapping specifically for 3d. The way voxel.html works, much like most Geometron apps is that it edits in real time the value of a file stored in the Geometron server at data/glyph3d.txt. This glyph is a string consisting of a sequence of base 8 numbers separated by commas as usual in Geometron. Also, as usual, as we edit a glyph the text input with the glyph code is updated live. We can always copy and paste any glyph from any instance of Geometron to any other. So if you are making a glyph in your local private server and send a private text message to another person with the text in the glyph input copied, they can paste that into the private server on their network and instantly replicate your glyph in their system. The way the app voxel.html works is that it has a local function in the HTML code which acts on the Geometron Hypercube by manipulating global geometric variables x, y and z, and constructing shapes in an x3d object. The specification of x3d contains basic geometry primitives such as cube and sphere, as well as transformations like translation and scale operations.

All of this work is carried out using the x3dom JavaScript library x3dom.js. This is documented at www.x3dom.org. This library allows for creation of x3d objects in an HTML file using DOM(Document Object Model) JavaScript com-

CHAPTER 13. GEOMETRON IN 3D AND BEYOND

mands. As the JavaScript code alters the x3d object, however, it creates actual human readable code which can be copied into a separate x3d file. This is done with the "save" button which copies the x3d code into another html file wrapped in a header and footer to make it render in a browser, called three.html. Every time someone hits the "save" button the old version of three.html is destroyed and replaced with a new version with whatever the current status of the 3d object is in voxel.html. The source code of three.html can be copy/pasted as a x3d object which can be embedded in any type of VR system which uses that format. This can be used to create games and virtual reality or augmented reality applications based on Geometron.

Another method for creating 3d graphics in the Web is the use of WebGL, a cross-browser graphics framework, via the JavaScript library Three.js. This fantastically powerful and well-documented library is another free open source library with publicly available CDNs for easy use. For more details see the home page of the project at threejs.org. The Geometron app threejs.html opens the same glyph file edited by voxel.html and renders it using three.js. This passive program serves two purposes. First of all, it implements the whole basic Geometron Hypercube using the three.js library, which can form a jumping off point for developers reading this to create more interesting applications such as games or other dynamic user interfaces. But its more immediate utility is the three.js library's ability to export to the .stl format. This format turns 3d surfaces into meshes of triangles, and is the universal language used by 3d printer software. Thus the ability to export from a Geometron glyph to .stl allows us to 3d print objects we create in Geometron. This adds yet another physical layer to the types of media which we can exchange using Geometron code. When the page threejs.html loads it automatically saves another file called data/three.stl, which can be downloaded by right clicking on the link, and that .stl file can be shared with anyone with a 3d printer to print it out. That .stl file can also be opened with a lot of 3d editing software and integrated into other projects, allowing us to make icons which other people can integrate into other physical designs for 3d printers or even CNC and/or injection molding. Note that the .stl files can get very large, and above a certain size the Geometron software will fail to save, as there is a limit to the size we can pass via the POST command in PHP. We leave this limit in place to keep the system working with modest sized files. This limit often makes spheres problematic and we replace them with cubes when we want to 3d print.

The app shapes3d.html allows us to edit Geometron glyphs stored in addresses from 0600 through 0607 in the Geometron Hypercube using the same editor style as in voxel.html. Up and down arrows cycle through the addresses. As glyphs are edited in each address, the global value of the hypercube stored at data/hypercube.txt is

CHAPTER 13. GEOMETRON IN 3D AND BEYOND

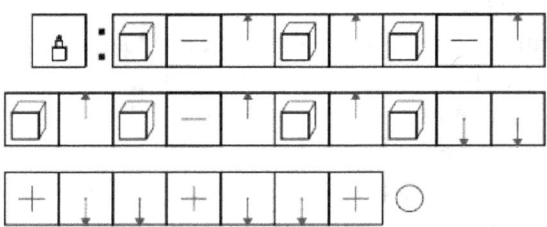

Figure 13.4: Turret construction. Cubes of decreasing size make a whimsical turret.

edited in real time. Each action sequence glyph in the range 0600 through 0607 has a corresponding symbol glyph made of 2d Geometron actions in the address range from 01600 through 01607. These are edited in a similar manner to other Geometron Hypercube stacks using the app shapes3dsymbols.html, which is linked from shapes3d.html.

As an example of using this method of building up actions to make shapes which are built up into more complex shapes, we will create a simple castle object. This starts with constructing a single tower made up of cubes of decreasing size(see figure). We can then repeat this turret and move repeatedly to build up a square wall to make a sort of castle, the Castle of Whimsy(see figures).

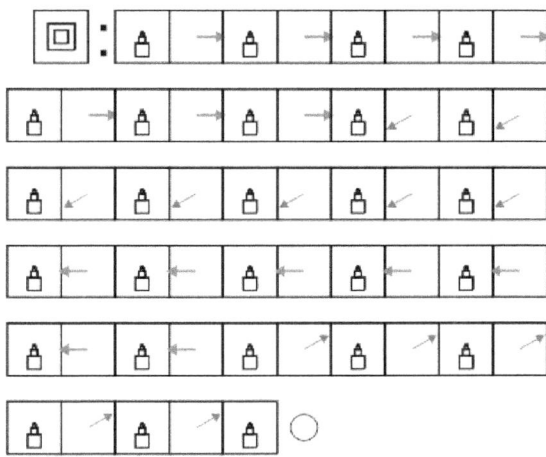

Figure 13.5: Many turret glyphs can be called to build up a castle.

We will now explore how the system described above is combined with the overall power of Trash Robot and Geometron to build symbols which link to the rest of the system here. The whole system of creating symbols for sharing in Trash Robot is based on pixel movements stored in addresses 0500 through 0507. These elements as previously documented both draw pixels on a screen in a canvas element(or SVG) and also control the motors in any of a number of robotic systems which prints in any of various materials. We now add yet another sequence

CHAPTER 13. GEOMETRON IN 3D AND BEYOND

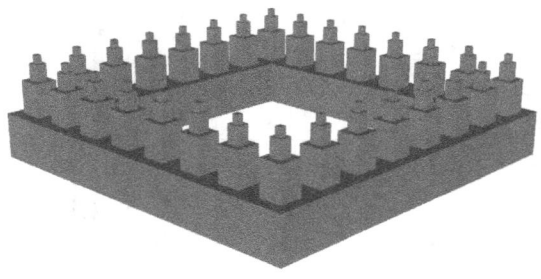

Figure 13.6: The finished product: a whimsical castle object, ready to use in VR, AR, or to send to a 3d printer.

of actions to these: movement with the drawing of a cube as a pixel(also known as a "voxel"). As with the robot control, We further add 3d geometry actions to the movement commands 0504 through 0507, so that we have the ability to move the 3d cursor around either with or without drawing a voxel. If we combine this with creating a base tablet in the form of a large array of cubes in a plane, we can create either 3d web assets for VR, AR and other 3d web graphics and also can create 3d printable objects to share, all from the same robot code used in the rest of the Trash Robot system. Just as we have an app for choosing an icon from the Trash Robot icon feed to print in the Trash Robot two and a half D printer, we have an app called icon3d.html which has a listing of all the Trash Robot icons, and when one is selected it loads up a 3d object with the icon printed in cubic voxels on the surface of a 8x8x1 rectangular solid. As in voxel.html, clicking on the "save" button will save the file to three.html, and loading the page threejs.html will resave the value of three.stl, overwriting the old version, for download and 3d printing. This system means that we can freely share feeds of icon glyphs which can be physically shared by 3d printing in addition to all the other methods of sharing.

3d printed parts can be used to stamp the symbol into soft objects, or used to make molds used to cast pourable objects with the icon in them. Simple modification of the values of the hypercube to reverse the sign of the x direction can create a mirror image in the 3d object, which can be useful if we wish to create a stamp which when imprinted into soft objects directly prints the icon without having to create another negative. Thus we can use 3d printing and basic Geometron operations to create copies of freely shared icons in any of a vast potential universe of materials both hard and soft. We can store feeds of icons in Geometron Trash Robot code in globally accessible text files, and link to those text files on pages on Geometron server domains which we link to in URLs printed directly on 3d printed parts. This makes the whole system self-replicating. Someone can pass you a 3d printed part which has an icon on it which directs you to a domain name which hosts a site which links to a feed which has a list of icons all of which can be replicated in another 3d printer just by clicking on them. This represents a decentralized network of free sharing of physical objects imprinted with icons which represent arbitrary things, to go along with all the other versions of this documented in other chapters.

Rotations and Higher Dimensions

As with every part of Geoemetron, we are only scratching the surface in this work of what can be done with these methods as we add actions to the Geometron Hypercube. Of the available 64 actions in the Hypercube reserved for 3d geometric actions we have so far only used 18. Any type of discrete geometric action can be added. In particular, rotations of rigid bodies is a deep field of mathematics which we have so far ignored. So far we have only moved our geometric axes around along the x, y and z axes or scaled them all together. We can add non-uniform scaling to make a much wider range of objects, and discrete rotations to create different orientations of objects. We can also add a larger range of geometric primitives, such as cylinders, cones, and other custom, manually-constructed polyhedra.

Discrete rotations can be a huge enabler of applications for Geometron. By constructing spheres, moving along a radius vector, rotating, and moving, we can build up three dimensional models of molecules. These

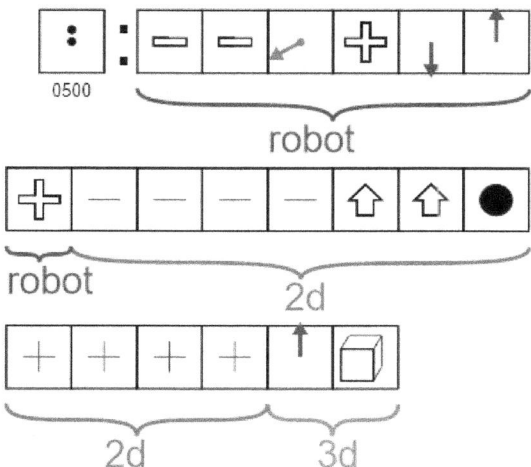

Figure 13.7: Breakdown of how the glyphs in the pixel code for constructing icons are built up by accessing different layers of the Geometron Hypercube. First we move the mechanical printing machine if there is one, then we create the little circle for the pixel on the screen if there is one, then create the cube pixel in the 3d web graphics object if there is one. Any parts which act on something that doesn't exist are ignored by the GVM, but the GVM abstraction allows one unified structure to describe behavior in all of these layers.

238 CHAPTER 13. GEOMETRON IN 3D AND BEYOND

Figure 13.8: An icon 3d object. This is from a 3d web file, but can be converted into an .stl file to print on a 3d printer exactly as shown, and then it can be replicated by using it as a stamp to create molds which are used to make more copies.

molecules can then be simulated by recasting problems of how they fold in terms of angular rotations of the Geometron glyph used to construct them. Sequences of chained discrete rotations can also be used to describe state vectors in the Hilbert space spanned by the qubits in a quantum processor. An alternative way to think about quantum computing can therefore be constructed in which instead of programming bit states, we use Geometron to express states in terms entirely of rotation angles, and all gate control pulses can then be thought of as further discrete rotation operations. If we map both states and dynamic operations in a quantum computer to Geometron glyphs, and also map problems of quantum chemistry(e.g. protein folding) into the same language, is seems at least plausible that such mapping could yield useful results in attempting to solve protein folding problems using the so-called NISQ(Noisy Intermediate Scale Quantum) systems which are presently available. Also direct mapping between geometric operations in graphics displayed in a web browser and states or operations on a quantum processor admits the possibility of a very simple and powerful real time user interface for development of quantum algorithms, again on existing NISQ devices. Such software could be easily created on top of the software frameworks currently under development by several parties. While this author believes quantum computing to be a complete waste of time which will never work, I find it amusing to pose this potential application in case someone wants to try it out, at least to make weird art on a quantum processor.

CHAPTER 13. GEOMETRON IN 3D AND BEYOND

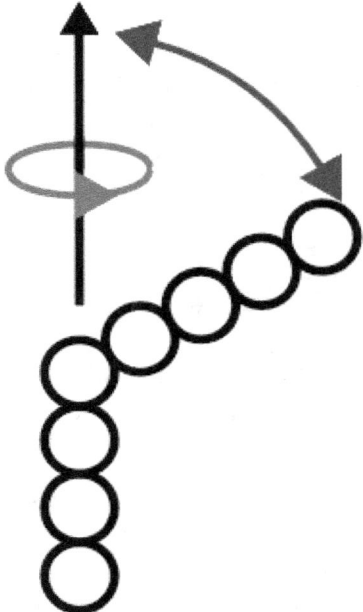

Figure 13.9: This shows how a sequence of objects can be constructed with each object being a discrete distance along an axis of specified altitude and azimuth. These angles can be manipulated with discrete rotations combined with step angle manipulations, just like the ones used in two dimensional Geometron constructions.

Chapter 14
Magic

Magic is self-replicating desire. When we desire a thing and that thing comes to pass, that is a form of replication from our minds into either the physical world or into the minds of others. This is the true purpose of Geometron: to create sets of self-replicating symbols in the most general sense discussed earlier in this book. Geometron could therefore also be called Symbol Magic, as it is a system of creating symbols themselves by replicating the desire to replicate them in other human minds.

Trash Magic defines our ultimate goal in this work: the creation of the Complete Set. A Trash Magic Complete Set is a self-replicating set which contains all the things we need as a community to live easy, comfortable lives with room for adventure in equilibrium with whatever ecosystem we are in. This means we have technology for sanitation, clean water, illumination, heating and air conditioning, food production, production of medications and medical technology, transportation and the self-replicating media of which these documents are the seed. And all of this technology must be built entirely from the waste we find in our immediate physical environment from the broken pieces of the old world combined with the natural flow of water, material, life, information and energy through our local ecosystem. Only by firmly fixing in our minds this end desire and building a new system of knowledge from scratch with these ends in mind, can we break free of the existing system which forces us all to do work we hate to buy things we don't want which destroy the Earth and hurt other people.

In order to build up this new way of seeing the world, we dive down to the deepest level of human thought, which is the subject of "things" in the most general sense. What is a "thing"? What is a collection of "things"? These are subjects mathematicians have tackled with vigor for a long time. At the end of the 1800s and for the first half of the 20th century, a collection of philosophers and mathematicians tried to build a foundation of mathematics from the theory of sets and logic. Their goal was to build something that was "true" in some deep sense which turned out to be elusive. In the views of this 20th century mathematicians created a linguistic trap in which mathematicians built increasingly meaningless worlds of symbols with no impact on the real world at all, leading to essentially an entire lost century of mathematics. The purpose of most academic mathematics is to gain recognition required for tenure, nothing more and nothing less.

Our goals here are much more practical. We want to build up whatever foundational ideas about language and mathematics most effectively aid us in building our new civilization from trash. This means we want to prioritize replication at every level of definition. In addition, we want to create our whole system based on trying to replicate the underlying desire to build up our new civilization. To this end, we ignore the axioms of 20th century "set theory", and define something we call "set magic". In Set Magic, we define sets to be collections of objects in the most general sense possible, just as mathematicians in the past did. But we note that for any given set, there are a vast number of choices as to how we break that thing down into constituent elements depending on what we want to do with our definition. In Set Magic, we always break things down into whatever collection of constituent elements are the most effective for communicating with other people what we need to communicate in order to replicate a thing. This means we choose sets with only a few elements, each of which can be summarized with a recognizable symbol, and which are related to each other with an easy to understand relationship. We define complex objects by building fractal structures of meaning, where things are made up of things, which are made up of things.

In Set Magic, we can assign an icon to any thing, as well as any element which makes up a thing. We can then use Action Geometry to make physical symbols which are easy to use to communicate meaning. We can call these "sigil boards", but they can also be thought of as a generalized game board. Geometric patterns can be drawn on cardboard with the Action Geometry sets and markers to make boards, and then we can use the clay Icon Tokens described in the Printers chapter to represent any set of objects in the most general possible sense. Each element can then be defined as another set, and another collection of self-replicating clay tokens can be created which are used to represent that set and so on. This can be thought of as a kind of physical symbolic hypertext, where self-replicating clay media(replicated via Geometron Printers) are arranged on self-replicating cardboard media(replicated via Action Geometry) to communicate any idea which can be parsed by the human mind to another person. Each thing can be a link to another set and so on. These self-replicating sets can be used to replicate anything. They are indeed used to replicate conceptually this whole system.

When we define sets with specific practical goals in mind, we find that the axioms of existing set theory are not satisfactory. The axiomatic set theory which has dominated for decades(so-called "ZFC") has a long list of axioms which apply restrictions on what can or cannot be an element or subset that are too limiting for our purposes. For example, we need to be able to have sets be elements of each other. For example, I might define myself to be a set of aspects of myself, one of which is my community, but of course a community is a set of which I am an element. While placing the set theorist inside the set they create is not strictly forbidden in old set theory, it is also not discussed. In Set Magic we explicitly put ourselves in the sets we create. We also put abstractions we can talk about in them, such as desires, ideas, concepts, symbolic archetypes and so on. Again, our mathematics has as a goal communication for replication for specific goals, and we will create any symbols and ideas which further those goals. A huge fraction of the ink spilled by set theorists has been in regards to their theory of infinite sets, which we also discard as mostly useless. Calculus lets us squeeze infinities down into symbols which we can work with and there is not need to waste our lives trying to build structure in sets with high order infinities which do nothing to help us communicate in the real world.

Another divergence in culture from classical mathematics is that we take geometry to be more fundamental than arithmetic. Mathematics in school is presented as the manipulation of numbers, with geometry as a secondary concern. In our mathematics, geometric constructions are the most fundamental thing. Geometry with meaning is a symbol, and symbols in their most general sense includes our entire technological system we are building, as well as all possible symbols to represent all possible human thoughts. Shifting from a world view in which numbers are fundamental to one where self-replicating geometric constructions are fundamental represents a deep shift in value system. Numbers are used to determine how owns what. They are used to buy and sell things, to track ownership, and to track people in systems of control. The ideology of numbers has taken over society as a sort of religious cult, centered on finance, computers, and authoritarian control of bureaucrats. When we focus our attention primarily on symbols which can replicate freely, it shifts our whole way of thinking and existing in the world. If I want a symbol to replicate, the easier it replicates the better. Numbers do not replicate, they are designed to help people control and dominate things in a world of consumption in which all people are in competition with one another. Put this in the simplest possible terms, numbers help people compete and dominate and symbols help people communicate and share. The elements of Geometron and the magic presented in this chapter represents an attempt to build a universal symbolic language which can form the basis of this new method of thought and of existence.

The Trash Robot is a self-replicating set which we create in order to grow into the system which can ultimately create Trash Magic Complete Sets. The Trash Robot set includes us, the people replicating the set, as part of replication is to grow the set by getting more people to join the set. It also includes our desire to build a better world as a thing(this desire can be represented in symbols therefore it can be an element of a set). The elements of the basic set are what are documented in this book. As with everything here, this has a fractal structure and can be parsed multiple ways into constituent elements. Trash Robot includes all the constructions described in this work, as well as the open brand defined by googly eyes, rainbow duct tape, things made from trash, bright colored felt on black cloth, and geometry. Join the set, become an element! And then help replicate the set. If we replicate the set together and replicate the intent to build the Trash Magic Complete Set, it should be possible to build that even if it is very difficult. From a fundamental technical standpoint, it is clear that building a Complete Set is possible, all we need is a community of people with the will to carry this out. Building the magic set with intent is therefore enough: if it replicates and evolves, it will create what we need, even if we have no idea how to do that.

The symbols we use to replicate our sets are chosen purely based on what works. Unlike the mathematical philosophers of the early 20th century, we are not seeking some higher objective "truth". Our entire theory of knowledge is just based on what symbols will, if shared, replicate our intent to build Trash Magic. Therefore we borrow from any existing symbolic framework that is convenient to use to our ends. This can include any philosophical, religious, spiritual, mythical, cultural

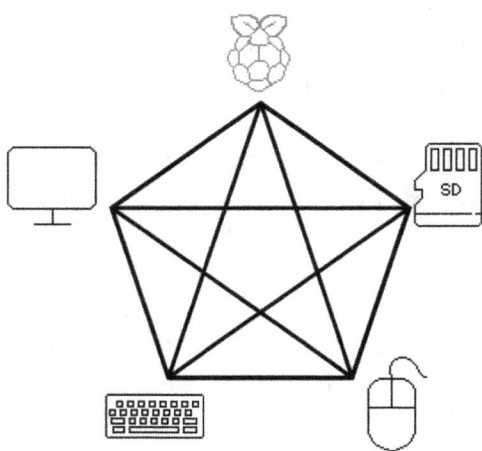

Figure 14.1: Sigil for Raspberry Pi. This figure was created with a Geometron Map combined with symbols made with the main symbol app combined with the icon feed to make the icons. A magician who wants to share this sigil with someone in the physical world can use these icon glyphs to print the icons in clay, replicate them and paint them to make tokens, then make a cardboard board with the symbol shown using Action Geometry. They can then place tokens on the board to communicate about the system. The same geometric information can be used to make self-replicating information on any Geometron server. As above, so below: there is always a version of our geometry on the servers and one made from physical media.

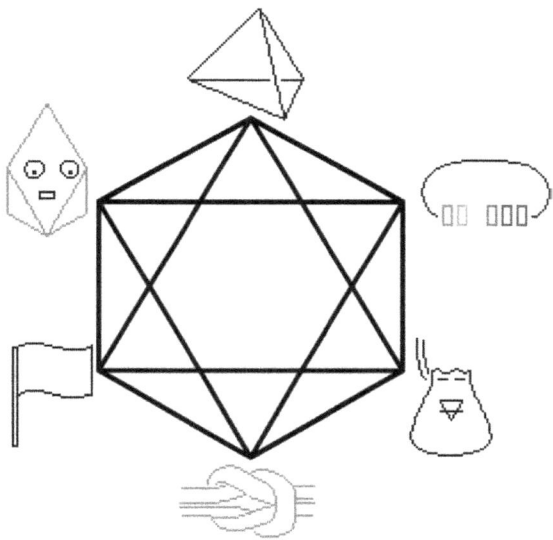

Figure 14.2: Sigil for Action Geometry. Action Geometry is a self-replicating set, which is itself an element of Trash Robot, Geometron, and any other set we find it useful to put it in. The layout of the sigil can be used to communicate relationships between things.

ideas, archetypes and symbols from anywhere. For various reasons, this work uses the archetypes of Alchemy frequently. The human mind finds it easy to deal with sets of five elements. And a pentagram in a pentagon connects each element to each other one. So breaking things into five elements, and mapping them to the five elements of Earth, Air, Water, Fire and Aether can be useful. This is not literal science! It is simply a symbolic mapping to aid in communication.

An example of the use of alchemy dividing the code in our system up into five languages: Water(HTML), Fire(JavaScript), Air(CSS), Earth(Geometron bytecode), and Aether(PHP). Also, the stages of creation of icon tokens are mapped to the elements with Air being the images, Aether being the glyph code for the icon, Earth being the print, Fire being the stamp, and the final colored in token being Water. The three kinds of clay pieces are stored in black cloth bags marked with the relevant alchemy symbols.

A word on notation borrowed from classical set theory is in order. We use the curly braces, also known as twiddle brackets, to enclose sets, which are lists of elements separated by commas. This notation also conveniently matches up with the JSON format used throughout the software elements of our system. We do not believe this to be a coincidence, and believe that JSON and HTML are both systems clearly built by people thinking along similar lines to the discussion here.

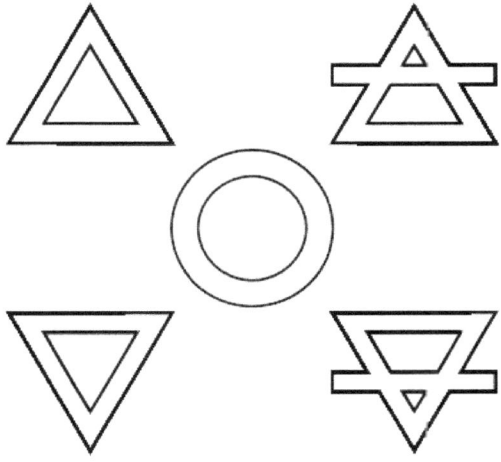

Figure 14.3: Symbols for the five elements of Alchemy. Aether is the circle in the center. Clockwise from upper left, the elements shown are Fire, Air, Earth, Water. Water is blue, Air is yellow, Earth is green or brown, Aether is purple, and Fire is red.

Figure 14.4: Sigil board constructed with Action Geometry.

Figure 14.5: Diagram a set.

Figure 14.6: Diagram another set.

Chapter 15
Full Stack Geometron

In spite of its philosophical intent, the media infrastructure presented here is all based on off-the-shelf information technology which ultimately comes from mines. We must state clearly, however, that our desire is to build a media system which requires no further input of mined materials to function. That is to say, all the media hardware we use should be based on waste streams from the existing system. This means it can be built on discarded and broken hardware, and then as that hardware continues to disintegrate over time, that it has the means to continue to repair and evolve it so that the constituent atoms and molecules and deep structures form the basis of a stable technology which can exist in equilibrium with the local ecosystem. To build this new type of media, we need to first be clear on what the goals are, then sketch out how this might be done, create a path forward for developing this, and then take action to set in motion the series of events which will cause this system to come into being.

When media does not have "users" and "data" as in consumer media, but is simply a network for hosting self-replicating documents, this radically alters some fundamental assumptions about how media works. Also, when we use the waste streams of the existing wasteful system, the basic mathematics of the situation is altered by the sheer astronomical scale at which waste media hardware is currently produced. If we combine the designed-to-fail phones with the designed-to-fail laptops, and tv's and so on, and then combine on top of that all the useless junk in the so-called "internet of things" which is currently being accelerated, the number of physical objects we can incorporate into our new media ecosystem is absolutely staggering. It is difficult to calculate the numbers and it is a moving target anyway, but it seems clear that we can easily build a system with at least a 10 to 1 ratio of recently-discarded media hardware to people, for every single person on this planet. That is to say, we can conservatively estimate that we have available a reservoir of 100 billion mostly-intact media devices to distribute among a projected population of 10 billion people. Less conservatively, and taking some of the marketing noise from Silicon Valley at face value, we might be looking at more like 100 or 1000 media devices per person, especially if we find ways to dig out a large number of more trashed older systems.

In a world with many media devices per person, and no private "user data" or property, we can build systems of ubiquitous community media rather than personal media devices(the primary function of which is to maximize waste, and to surveil and control people). What does this look like? To get a feel for how this could look we recall the discussion earlier in this book about the Street Network, and the power of physical places. Places with significance, such as cross roads, major cultural centers, community centers, parks, fords, bridges, tunnels, libraries and so on can all host ubiquitous organic media.

If I want a document in a place other than where I was reading and editing it before, I simply replicate that document from one network node to the next. Documents can flow from one place to another, along with people. But the physical infrastructure of the media does *not* need to be moved from one place to another. This is a very radical departure from existing media systems, and it alters our design constraints on hardware dramatically. Perhaps the two most powerful effects of this difference on design is on portability and redundancy. If we are using streams of trash to build things, and are accumulating this material in central physical network nodes, we can have a very high level of hardware redundancy. In mine based systems, any redundancy means more money and more mining and is avoided at all cost, with the smallest amount of material in the smallest space. When devices are all attached to individual persons, they must be as small as possible, and less material is also always sought out. But when we are absorbing material which was a liability to society, the more of it we absorb the better. So we can build whole structures of stitched together screens for purely aesthetic purposes which never would have made sense in the old system. Also, we can build systems that just use a single component from something and let the rest of it sit there waiting to be absorbed into some future project, but unused for the time being.

Removing portability as a design goal makes a lot of design tasks much easier. Perhaps the most powerful consequence of this shift is with electrical power. Today's media devices all either take power from a wall socket or from a battery which is charged via a wall socket. Batteries are always pushed right up to the physical limits of light weight and high energy density. But if we replace this with static media infrastructure, batteries can be replaced by hybrid power cells which store energy by several means, and take up as much space as the materials used happen to need. We need a system for converting all the materials in wasted batteries today into working batteries. This is not "recycling" in the existing centralized sense, with mass production mirroring the process of mine-based production. Like everything in our new civilization, it is a craft mode of production, in which anyone anywhere in the world can use skills transmitted via the network to directly convert waste batteries into cells in a modular power pack which can be used to run all of our electrical devices. Again, we note that there is no reason based on the fundamental physics and chemistry of the system that this is not possible or indeed even easy. It is just not done today due to the broken system based on money, mining and property we are forced to work with in modern society.

In addition to re-designing the whole power storage system of our media, we need to address power generation itself. In order to eliminate mining and the control of empires, we need power generation which is based entirely on trash, and is totally local. The proposed civilization here will use less than 1 100th of the power of existing civilization by removing wasted efforts. *Most* power today is used to do pointless activities, from driving around in circles to keeping lights on in giant buildings where people do pointless things all day. A huge fraction of this is also based on our system of property which forces people to ignore weather, seasons, and natural geography and live places that make no sense to live based on the violence of the state and property. If we ignore this and build a new civilization not tied to property, much of the energy we waste on huge megacities in unsustainable locations can be terminated as we abandon those cities(Phoenix Arizona should probably have a population of under 1000 people). Micropower can be generated by converting all the motors found in broken electrical devices into power generators, and used for direct mechanical power generation, from wind, water, and heat engines run on the sun. Photovoltaics and wind as they are used today are not of interest, as they rely on mine waste just like coal, nuclear and oil. No mining means no mining. And that means we will only build where we can get some power from the natural world above ground.

Another consequence of non-portable media is that we can scale the displays way up. In the chapter of this book on Action Geometry, we describe a Trash Camp in which large modular structures are built for both shelters and industrial production infrastructure. Trash camps like this which accumulate a large amount of material can have essentially all surfaces turn into screens as wasted screens pile up from the old system. Some screens might even be dead, and just used as a flat structural element which blocks rain water. But if we can build a modular system for integrating screens into our media, and also build out the optics technology to do large scale projection on walls and other large surfaces, we can have truly ubiquitous displays if we want to.

In general, we also are always focusing on increasing modularity so that components which are repurposed from old broken technology have the maximum possible utility. In many cases, this will mean sacrificing the *majority* of the elements of a piece of technology and just making use of one component. We can then, over time,

Figure 15.1: Geometron Station. Sketch of a Geometron Station in a Trash Camp, showing what ubiquitous organic media might look like.

as our technology improves, find ways to repeat this with other components. But initially we will find that in many cases something like a whole smart phone is being converted to just a single modular device, say an accelerometer or a touch sensor. In order to facilitate this kind of modular technology, we need to develop a hybrid interconnect technology in which we build up fabrication at the millimeter scale, constructing again by craft production components which break out the electrical contacts in a simple, well-documented way from needed components to the outside world. In some cases we can build these interconnects on the existing standards like USB, but in many cases these standards are designed to make low quality stuff that fails quickly and we will need to spend considerable research efforts on building more robust standards. Much more robust standards will be possible when we again use the principles that larger size and redundancy are now acceptable. Without minimization of material use and size, contacts can be designed which are much more robust than the aggressively miniaturized connectors used by consumer technology.

When our network is primarily used to share documents, rather than real-time surveillance, manipulation and control, it also opens up some new possibilities in terms of information transport protocols. The Street Network is going to create a network of travelers on the physical networks of the world: truckers, hitchhikers, camper dwellers, and wanderers of all sorts. If we don't need documents moved instantly, this physical network can be used to transport physical storage media, creating a flow of information comparable to the trunk lines used to run the Internet today. This network protocol is a social one rather than a technical one. Rather than information in packets being directed to numerical addresses of machines, we have documents being sent from one community of actual people to another. This makes the protocol sound like mail, but it is not. Mail is still based on the unique address of an individual person or organization. The network we are building is not based on property or individuals, but on communities and on self-replication of information. Documents will flow to where they are wanted, and actual humans will make choices on the fly about what that looks like.

In building a new type of media infrastructure, we must also consider the fundamental issues of what our machines are designed to do. These are not computers. We are building machines to edit, replicate, and read *documents*. The fundamental operation of a "computer" is arithmetic. The fundamental operation of our machines is the display of symbols in the generalized sense discussed in an earlier chapter: any geometry with meaning. If the only purpose of a machine is to display a document and interact with it, the whole hardware and software architecture can be completely different. The toy model of a computer called a Turing Machine. which does things to numbers based on a program written in numbers is replaced by a model where a human makes the choice to engage a machine, requesting that the machine draw a symbol, which it does based on a program made up of geometric actions and symbols. This is not a computer! It is Geometron: machines using symbols to make symbols rather than numbers to make numbers.

In computers, a program is running all the time, and it carries out an operation on each clock cycle, with well over a billion clock cycles per second, even when it does nothing. In our machines, no permanent clock is needed. When we engage with it by clicking, scrolling, typing or engaging some other sensor, we are carrying out a geometric action which causes a glyph stored in memory to be drawn. This can have any kind of time steps, including uneven ones, where we simply want things to be done in a certain order, and each step takes as long as it takes to do the geometry. In some cases we will want things to happen fast, but in general we aim to make these things as simple as possible, so that the things happening on a nanosecond time scale only relate to saving and loading large documents to and from memory. When we do nothing, the machine is also at rest. And it only comes alive and does things in direct response to our engagement. All software which is not creating a document, editing one, or replicating one is dispensed with.

The Geometron language can be used to create any of the documents described here, down to the hardware level. We will have bitmaps be stored in an Image Stack, which is memory for the sole purpose of storing sequences of bitmaps. A Geometron Virtual Machine can then be implemented in hardware by translating sequences of addresses into geometric actions. These addresses reference a Geometron Hypercube which is a physical memory component which stores address sequences in each address. With geometric movements as well as a whole font stored in the font section of the Hypercube, scrolls can be constructed as just another Geometron glyph. Reading a book or article or any other kind of rich text document is then just a matter of drawing a single big Geometron glyph. This glyph can include layout, graphics, text, and complex rich text structures in multiple languages. Geometric constructions combined with access of the aforementioned Image Stack can be used to construct Maps on our displays. If we can display Maps, Scrolls, and generic symbols, we can create and edit any document, making a generalized symbol machine in analog to the generalized computing machine.

Information is stored in self-replicating media as described in the Printers chapter: lithography in brass to make stamped copies in removable silicone films. This information is both written and read out using old DVD drives for both the laser, the optics, and the positional control(at about 1 micron to a few cm scale). as well as the spinner to spin on the silicone coatings and photoresist layers on the polished brass plates. Information is encoded using the Roctal coding scheme(octal that rocks). In this scheme, numbers are represented in binary, with rows of three bits so that each row represents a digit in a base 8 system. These arrays of bits are arranged in some standard repeating pattern with added structures for alignment marks. This method encodes bytecode which can be used to create any of the documents described here: HTML files, JSON data, Scrolls, Maps, Feeds, Image Stacks(in base 64 encoding), and Geometron Hypercubes. Hardware which converts these types of information into symbols can be created as a physical implementation of the GVM, directly translating sequences of addresses into sequences of geometric actions either on a display or on some physical machine. The most fundamental such machine this controls of course is the writer. The writer writes code onto the brass plate which represents documents, including the code to write the code. This plate, after being etched, is used to replicate the information many times over using soft silicone films with imprints of the brass plate. These are then read by other people on other Geometron machines, so that another person can read a set of documents which describe how to build this whole system from trash other people can find in their physical location. Having replicated this media system, the next person who joins can then use their machines to build more self-replicating documents, by printing out brass plates and then replicating the documents again and again with one plate. Furthermore, each instance of the system can be printing in clay at a human readable size, and we can carry out physical tabletop symbolic interactions by moving printed clay tokens around with other people. As discussed before, these tokens are themselves self-replicating media, which can be used to stamp clay to stamp more clay and so on, replicating without a printer. And, like the microscopic brass printed information, these tokens which represent elements of sets are used to replicate the media system by containing all the elements of the system we need to discuss when talking to someone face to face about replication.

Numerous challenging details are left out of the discussion in this chapter, but that doesn't matter. What is clear now is that the described system can be built from a physics and engineering standpoint. If we want it, we can build it. So what is needed now is only to decide that we want it and that we know how we will exist in relation to it as it comes into being. If we can do that, and if we can share this desire widely enough, it all of the technical problems will be solved by someone. This is the full power of Trash Magic: we seek to replicate the desire to build everything from trash in both the minds of other people and in the physical reality of the world around us. If you join in this mission, and you fully commit to feeling the desire for this world, you should be able to share that with someone else, and if they can do that, they can pass it on. If we can spread this desire widely enough, the materials, skills, energy and time will simply materialize from the mass efforts of all these people. Building a new civilization without money, without mining and without property is a lot easier than you think. Now is the time for us to simply decide to do this, and get on with living the adventures that this new world will bring about.

www.ingramcontent.com/pod-product-compliance
Lightning Source LLC
Chambersburg PA
CBHW060825220526
45466CB00003B/978